SpringerBriefs in Economics

SpringerBriefs present concise summaries of cutting-edge research and practical applications across a wide spectrum of fields. Featuring compact volumes of 50 to 125 pages, the series covers a range of content from professional to academic. Typical topics might include:

- A timely report of state-of-the art analytical techniques
- A bridge between new research results, as published in journal articles, and a contextual literature review
- A snapshot of a hot or emerging topic
- An in-depth case study or clinical example
- A presentation of core concepts that students must understand in order to make independent contributions

SpringerBriefs in Economics showcase emerging theory, empirical research, and practical application in microeconomics, macroeconomics, economic policy, public finance, econometrics, regional science, and related fields, from a global author community.

Briefs are characterized by fast, global electronic dissemination, standard publishing contracts, standardized manuscript preparation and formatting guidelines, and expedited production schedules.

Jung Kyu Canci · Philipp Mekler · Gang Mu
Editors

Quantitative Models in Life Science Business

From Value Creation to Business Processes

 Springer

Editors
Jung Kyu Canci
Lucerne University of Applied Sciences
and Arts
Horw, Switzerland

Philipp Mekler
Roche (Switzerland)
Kaiseraugst, Switzerland

Gang Mu
University of Zurich
Zurich, Switzerland

ISSN 2191-5504 ISSN 2191-5512 (electronic)
SpringerBriefs in Economics
ISBN 978-3-031-11813-5 ISBN 978-3-031-11814-2 (eBook)
https://doi.org/10.1007/978-3-031-11814-2

© The Editor(s) (if applicable) and The Author(s) 2023. This book is an open access publication.
Open Access This book is licensed under the terms of the Creative Commons Attribution 4.0 International License (http://creativecommons.org/licenses/by/4.0/), which permits use, sharing, adaptation, distribution and reproduction in any medium or format, as long as you give appropriate credit to the original author(s) and the source, provide a link to the Creative Commons license and indicate if changes were made.
The images or other third party material in this book are included in the book's Creative Commons license, unless indicated otherwise in a credit line to the material. If material is not included in the book's Creative Commons license and your intended use is not permitted by statutory regulation or exceeds the permitted use, you will need to obtain permission directly from the copyright holder.
The use of general descriptive names, registered names, trademarks, service marks, etc. in this publication does not imply, even in the absence of a specific statement, that such names are exempt from the relevant protective laws and regulations and therefore free for general use.
The publisher, the authors, and the editors are safe to assume that the advice and information in this book are believed to be true and accurate at the date of publication. Neither the publisher nor the authors or the editors give a warranty, expressed or implied, with respect to the material contained herein or for any errors or omissions that may have been made. The publisher remains neutral with regard to jurisdictional claims in published maps and institutional affiliations.

This Springer imprint is published by the registered company Springer Nature Switzerland AG
The registered company address is: Gewerbestrasse 11, 6330 Cham, Switzerland

Preface

There has been a surge of interest in understanding business processes in healthcare over the past few years. While there are many treatises on particular technical and economic facets of such processes, little has been written on their specific quantification. Hence, the editors felt there had never been a better time to write a book giving an overview on quantitative aspects of relevant business processes in healthcare. As they cover a very broad spectrum of disciplines including mathematics, game theory, social sciences, machine learning, and economics, the diversity of topics could not be comprehensively covered in a single volume. However, this book aims to present a selective coverage of the core elements and recent topics from within the broad field of healthcare business processes.

This book is primarily directed at practitioners in the field of healthcare economics, applied probability, statistics, and machine learning. By keeping the mathematical prerequisites simple and to a minimum, the book will be of interest and accessible to the majority of readers. As far as possible, the development is self-contained while necessarily condensed.

The book is divided into three distinct parts, together with supplementary materials. The first part looks at *Value Creation and Managing Intellectual Property in the Life Science Industry*, an extremely important aspect of any knowledge-based industry. The second part *Modelling Specific Business Processes in the Life Science Industry* relates to some selected topics encountered in healthcare business. The third part *Specialized Quantitative Tools in the Life Science Industry* finally provides an insight on the development of some mathematical tools from analysis and probability theory.

All chapters discuss basic elements of business issues encountered in healthcare, ranging from fundamental properties to simulation methods, as well as discussing legal and societal ramifications. The appendix provides background material and code details as specified in the respective chapters.

Lucerne, Switzerland	Jung Kyu Canci
Basel, Switzerland	Philipp Mekler
Zurich, Switzerland	Gang Mu
March 2022	

Acknowledgments

First we would like to thank all the authors we had invited to contribute in making this book. Without them, their particular views, experience and knowledge, this book would not have been possible. We would also like to acknowledge their patience and magnanimity towards an often prodding editorial team. It took a lot of discussions, agreeing and disagreeing to take this from an initial idea to a final book. We hope the results justify the path taken.

In addition, our special thanks go to Emanuelle Mekler and Sai Kumar, our editorial support team out of F. Hoffmann-La Roche AG, Basel. Without them, we could not have met the many deadlines, the individual author contacts and the final product we see here. Their efficiency was instrumental, their patience endless. Many thanks, Emanuelle and Sai.

Additional thanks go to our and all authors' respective institutions and companies: Basel University, Lucerne University of Applied Sciences, National University of Singapore, University of Bern, University of Calabria, University of L'Aquila, University of Udine, and F. Hoffmann-La Roche AG, Basel. They have supported this book and all authors with resources of time, funds and expertise, without which this book would not have been written.

March 2022

Jung Kyu Canci
Philipp Mekler
Gang Mu

Contents

Part I Value Creation and Managing Intellectual Property in the Life Science Industry

Value Creation, Valuation and Business Models
in the Pharmaceutical Sector 3
Michael Blankenagel, Jung Kyu Canci, and Philipp Mekler

Limited Commercial Licensing Strategies: A Piecewise
Deterministic Differential Game 17
Domenico De Giovanni and Jung Kyu Canci

Partnership Models for R&D in the Pharmaceutical Industry 29
Gianpaolo Iazzolino and Rita Bozzo

Part II Modelling Specific Business Processes in the Life Science Industry

Pharma Tender Processes: Modeling Auction Outcomes 51
Philipp Mekler and Jingshu Sun

Multi-Echelon Inventory Optimization Using Deep Reinforcement
Learning 73
Patric Hammler, Nicolas Riesterer, Gang Mu, and Torsten Braun

Part III Specialized Quantitative Tools in the Life Science Industry

An Invitation to Stochastic Differential Equations in Healthcare 97
Dimitri Breda, Jung Kyu Canci, and Raffaele D'Ambrosio

Life Events that Cascade: An Excursion into DALY Computations 111
Young Lee, Thanh Vinh Vo, Derek Ni, and Gang Mu

Value Creation and Managing Intellectual Property in the Life Science Industry

Value Creation, Valuation and Business Models in the Pharmaceutical Sector

Michael Blankenagel, Jung Kyu Canci, and Philipp Mekler

1 Value Principles in the Pharmaceutical Industry

The pharmaceutical industry is a key, yet complex sector within the global economy. Organizationally, its complexity is outlined by an involved business model, an intricate organizational structure, and a challenging environment. Economically, the pharmaceutical industry has been characterized by high profit margins; this mainly as a result of substantial research and development (R&D) investment and its legal protection by patents. Over time the original situation has evolved further, generating two major types of pharmaceutical firms: originators and generic producers. High R&D investment is a characteristic of the originator pharmaceutical companies which produce patent-protected drugs, as well as biotech specialists which produce biologics. The generic producers, on the other hand, do not incur the initial R&D expenses (or less so) and in general produce drugs lacking patent protection. On top of this now traditional set, new segments have arisen in the pharmaceutical industry, comprising services in or around the traditional drug industry, e.g. diagnostic or data-oriented endeavours.

What defines the process of value creation in pharmaceutical firms? In the long run, it is the role of successful R&D as a driver of value creation. This long-term view of value creation has particular implications: (i) R&D is a critical input to long-term

---

M. Blankenagel · J. K. Canci
Lucerne University of Applied Sciences and Arts, Lucerne, Switzerland
e-mail: michael.blankenagel@hslu.ch

J. K. Canci
e-mail: jungkyu.canci@hslu.ch

P. Mekler (✉)
University of Basel, Basel, Switzerland
e-mail: philipp.mekler@unibas.ch

© The Author(s) 2023
J. K. Canci et al. (eds.), *Quantitative Models in Life Science Business*,
SpringerBriefs in Economics, https://doi.org/10.1007/978-3-031-11814-2_1

growth and the pharmaceutical sector is one of the highest R&D-intense sectors, (ii) this intense R&D effort is only economically feasible when protected by intellectual property legislation and (iii) successful R&D leading to the discovery of new drugs increases its economic footprint by improving the society's health status and well-being. The present chapter attempts to outline value creation, value protection and value estimation using the above ideas.

2 Value Creation in the Pharmaceutical Industry

In the pharmaceutical industry value is typically created in one of four business modalities: (1) disease solution providers, (2) breakthrough innovators, (3) commercial optimizers and (4) value players (Behnke et al. 2014; Buldyrev et al. 2020; Clark et al. 2021).

(1) Disease solution providers:
   Such companies approach competition by offering differentiated products and services based on thorough understanding of the disease and customers. Gilead's unique HIV combination therapies drove an eightfold increase in the company's share of the HIV/AIDS drug market in the 2010s. As another example, Novo Nordisk's leadership in diabetes care largely explains why its 2016-20 EBITDA margin was higher than would have been expected from its relative share of the pharma market as a whole.

(2) Breakthrough innovators:
   Such companies create one-of-a-kind products, requiring less emphasis on sophisticated commercial capabilities. For example, around 2010, Celgene (since 2019 a Bristol Myers Squibb company) changed the game in multiple myeloma by developing innovative applications for the historically negatively connotated Gruenenthal drug thalidomide. Roche built its leadership position in oncology on Genentech's breakthrough work in systematically developing humanized monoclonal antibodies.

(3) Commercial optimizers:
   These extract maximum value from proven, not always highly differentiated, products. A typical example is Pfizer, which built a dominant position in the branded primary care category by figuring out how to commercialize acquired assets, especially products that lacked significant clinical differentiation.

(4) Value players:
   These are companies having achieved leadership in generics by deploying differentiated business capabilities to build scale and breadth in their target geographies. Such companies achieve success by developing differentiated business capabilities; India-based Cipla or Teva out of Israel may serve as typical examples. Cipla has focused on manufacturing low-cost generic drugs for fatal diseases afflicting large populations in developing countries. Teva has succeeded in the US and other Western markets by successfully challenging the intellectual

property positions of originator companies and being first to market with new generics.

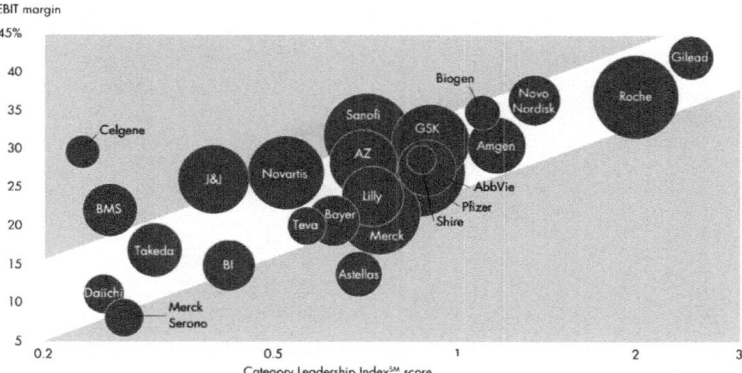

Illustration: Category leadership versus profitability (adapted from Behnke et al. (2014)).

3 'Keeping Focus': The Traditional Value Token

Since the early 2000s building leadership in a particular value creation category has become crucial for success in pharma. Seven of ten leading value creators, e.g. Roche in oncology and Novo Nordisk in diabetes care, generated at least 50% of their revenues from one particular therapeutic area. In some extreme cases (e.g. Biogen in neurology and Incyte in oncology) more than 90% of revenues came from a single therapeutic area.

Category leaders have privileged access to all stakeholders in a given category. This allows them to identify and satisfy unmet customer needs, often at the intersection of science, logistics and marketing. Their product and regulatory functions benefit from more expertise and stronger relationships, enabling them to get innovations to market faster and with a higher success rate. They are well placed to understand and price the best business development opportunities and are a preferred partner for smaller companies to develop and market their products. Lastly, their market presence and strong customer relationships improve commercial efficiency.

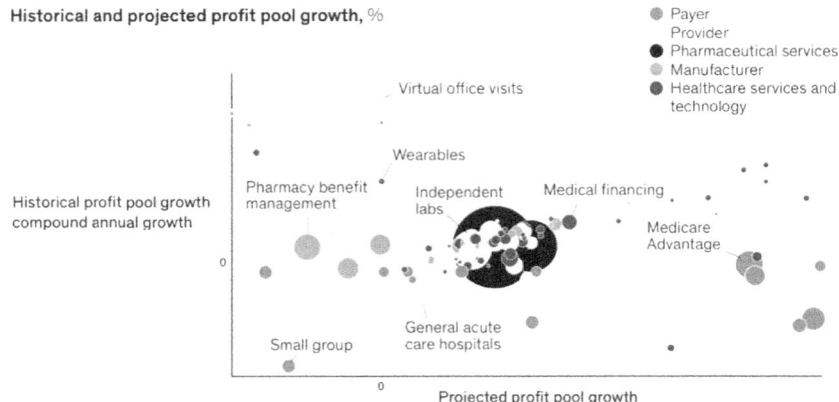

Illustration: Profit growth by business area (adapted from Clark et al. (2021)).

4 'Extending Horizons': Innovation-Integration Across the Value Chain

Outside of classical pharma, growth in healthcare services and technology has been accentuated, as old and new players are bringing technology-enabled services to help improve patient care and therapeutic efficiency (Clark et al. 2021). Healthcare services and technology companies are serving nearly all segments of the healthcare ecosystem. These efforts include working with payers and providers to better enable the link between actions and outcomes, to engage with consumers, and to provide real-time and convenient access to health information. Venture capital and private equity have fueled much of the innovation in the space: more than 80 percent of deal volume has come from these institutional investors, while more traditional strategic players have focused on scaling such innovations and integrating them into their core. Driven by this investment, multiple new models, players and approaches are emerging across various sub-segments of the technology and services space, driving both innovation (measured by the number of venture capital deals as a percent of total deals) and integration (measured by strategic dollars invested as a percent of total dollars) with traditional payers and providers. In some sub-segments, such as data and analytics, utilization management, provider enablement, network management and clinical information systems, there has been a high rate of both innovation and integration. For instance, in the data and analytics sub-segment, areas such as behavioural health and social determinants of health have driven innovation, while payer and provider investment in at-scale data and analytics platforms has driven deeper integration with existing core platforms. Other sub-segments, such as patient engagement and population health management, have exhibited high innovation but lower integration. Traditional players have an opportunity to integrate innovative new technologies and offerings to transform and modernize their existing business models. Simultaneously, new (and often non-traditional) players are well positioned to continue to drive innovation across multiple sub-segments and through combinations of capabilities.

5 Value Protection: Intellectual Property in the Life Sciences

In his paper on business innovation and growth (Ahlstrom 2010), David Ahlstrom argues that the main goal of any business is to develop new and innovative goods and services that generate economic growth while delivering important benefits to society. Steady economic growth generated through innovation plays a major role in producing increases in per capita income. Small changes in economic growth can yield very large differences in income over time, making firm growth particularly salient to societies. In addition to providing growth, innovative firms can supply important goods and services to consumers.

Classically, among the more advanced methodologies, static net asset value (NAV)-based valuations have been used to attempt catching the 'true' value of a patent. However, it has become increasingly evident that uncertainty in a patent's life cycle must be considered when performing patent valuation. For these reasons, a new family of quantitative models which account for uncertainty by means of stochastic (Monte Carlo) simulations have been used by several groups and companies.

5.1 Patent Evaluation

5.1.1 General

A key feature of patents in the pharma and biotech industries is that their value is uncertain. There is a large gap between patent value studies and cost-benefit analysis tools. Existing valuation approaches do not consider a patent's life cycle, an important and unique characteristic of pharma and biotech patents.

Hence, some authors propose a quantitative stochastic model that accounts for uncertainty and solves the problem by means of Monte Carlo simulations. This is done to model the uncertainty in a patent's value as a stochastic process and use a mean-reverting process to model changes in the value during the patent's life cycle. Furthermore, one can perform comparative parameter analyses and discuss the implications of the proposed model.

5.1.2 Pharmaceutical Patent Evaluation Approaches

As exemplified by Banerjee et al. (2019), one can classify typical patent valuation approaches into two different groups: an expert approach and a monetary approach. The most intuitive approach is based on expert knowledge, which can be considered easy, often proprietary, and sometimes quite subjective. It mainly relies on comparison metrics, sum-of-parts values and on historical precedents.

The monetary approach, on the other hand, tries to evaluate the patent's economic value via monetary categories such as cash flow or profit patents may be able to generate in the future. These methods can be further sub-grouped on the basis of their operating approaches: (1) the cost approach, (2) the market approach and (3) the income approach.

(1) Cost approach:
   In this approach, patents are valued on the basis of reproduction cost (i.e. all cost associated with purchase or development of a replica of patent under consideration) and replacement cost (i.e. cost to be incurred to obtain an equivalent patent asset having similar use/or function). In both of these methods, the present prices are considered. Typical heads include cost of research and development, promotional expenses, management time, legal licensing and registration fees, and opportunity cost (if any). The method also takes into account obsolescence costs like technological, economical and functional obsolescence.
(2) Market approach:
   In this subgroup, the patent value is estimated by taking reference of open market values, where there is evidence of prices, at which similar assets with similar uses have changed hands. If the asset is unique in nature, then comparison is done on the basis of utility, technological specificity and property. Data is collected from different sources like company annual reports, specialized database of royalty rates, stock price, legal decisions and pure patent deals.
(3) (Mixed) income approach:
   Under this approach, the patent is valued on the basis of the future benefits that would accrue from the concerned patent and discounted by an appropriate discount rate. Often such models of patent valuation have been obtained from the academic literature. These can be categorized into four sub-groups, i.e. income approach, indicator-based approach, mixed approach and market approach according to their working methodology. The most pertinent (mixed) income approaches are tabulated in Table 1.

The approach based on net present value (NPV) is well accepted, but static. Here, the NPV of a patent is derived by comparing all expected future cash flows generated by the patent with the expected costs to determine whether the patent will be profitable. A positive NPV suggests that the patent will be profitable. NPV is the dominant patent evaluation approach, but limited because of static future revenues assumption. Some adjustments of NPV have been proposed (risk-adjusted NPV, using different interest rates to more or less discount future revenues), and still do

Table 1 Comparison of (mixed) income approach methods of patent valuation (adapted from Banerjee et al. 2019)

Source	Methodology	Advantages	Disadvantages
Reitzig (2000)	Option	Several risk effect factors incl.	No asset risk change over time
Leone and Orianim (2021)	Option	Patent option characteristics	No asset risk change over time
Triest and Vis (2007)	Income; DCF	Economic patent value	Needs market/technology info
Sebastian et al. (2010)	Option; Simulation	Project time risk incl.	Assumption restricted model
Meeks and Eldering (2010)	DCF method	Technology and litigation data	Needs historical transaction data
Sereno, 2010	Option based DCF	Values tech./proc. innovation	No asset risk change over time
Sohn et al. (2013)	Classification tree	Willing-to-sell/buy angles	Practice constraints
Russel (2016)	DCF; Value weight	Investor valuation disclosed	Needs CF, disc.rate, expiry data

not account for uncertainty explicitly. This always assumes that future cash flows will be fixed.

5.1.3 Patents as Options

To view patents as a volatile financial asset, elements out of option pricing theory have been used. Here, in contrast to the traditional NPV approach, real option theory provides a more realistic way to value strategic growth opportunities and uncertainty. In addition, decision tree method to value a biotech company based on its R&D (Kellogg and Charnes 2000) is being considered, as well as an abandon-option view when valuing patents and patent-protected R&D projects (Schwartz 2004). The underlying uncertainty view is critical for valuing patents and because the dynamic characteristics of patent value are inherited. Combining real options with binomial trees to assess patent renewal strategies has also been studied (Baudry and Dumont 2006).

5.1.4 Patent Evaluation Using the S-Curve Life Cycle

The completion of successful pharmaceutical R&D steps in each phase increases the potential value of a patent. In the early stages of patent licensing, the patent's value is low due to risks and uncertainties. Later (phases 1–3) the value grows as the potentially huge market revenues protected by the patent are realized. This underlines the importance of considering life cycles in evaluating pharmaceutical R&D programs: different phases of drug R&D generate diverse risks (Myers and Howe 1997). Risks that have a significant effect on a patent's value gradually diminish over time until the final market launch phase is reached. The patent value changes dramatically during the life of an R&D project; the company needs to adjust its cash flow in different phases of R&D (Villiger and Bogdan 2005).

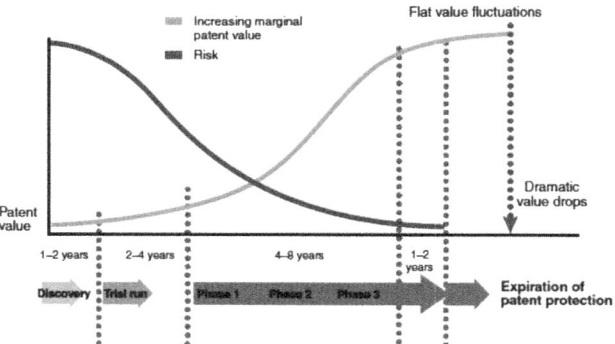

The S-shape curve life cycle of patent value (adapted from Wu and Wu (2011)).

6 Modelling the Patent Value as a Stochastic Process

The patent life cycle is modelled as a standard stochastic mean-reverting process (Ornstein-Uhlenbeck mean-reverting).

6.1 The Patent Life Model

- To describe the dynamics of the patent value V as a stochastic process, assume that V follows the standard Brownian motion.

$$\frac{dV_t}{V_t} = \alpha_t dt + \sigma_1 dz_1.$$

- This indicates that the patent value V is uncertain and stochastic over time. The instant rate $\left(\frac{dV_t}{V_t}\right)$, the change in V, accounts for two sources of uncertainty.
- The drift α_t represents the slope of the long-term path of V. The second term (σ_1) characterizes the volatility of the patent value process, where dz is an increment of a standard Brownian motion.

- Applying the life cycle of the patent, value stochastically converges the initial high growth rate generated by patents to a reasonable and sustainable growth rate over time. To ensure convergence of the drift α_t, it is modelled s.t. the slope of the long-term path of V follows a mean reversion process:

$$\frac{d\alpha_t}{\alpha_t} = \eta(\bar{\alpha} - \alpha)_t dt + \sigma_2 dz_2.$$

- This denotes a standard uncertain process for the drift α_t, (Ornstein-Uhlenbeck process); η represents one half of the decay rate of the drift α_t, which moves the long-term average drift $\bar{\alpha}$. The change in the speed of adjustment $\eta > 0$ measures the mean compared to the mean drift.
- The above equation is a continuous expression of the patent value V, and the patent value under uncertainty is simulated by converting it into a discrete form expression.
- By Ito's lemma the equations shown above can be re-written as follows:

$$V_t = V_{t-1} e^{(\alpha_t - \frac{1}{2}\sigma^2)\Delta t + \sigma_1 \varepsilon \sqrt{\Delta t}},$$

where $\alpha_t = \alpha_{t-1} e^{-\eta \Delta t} + \left(1 - e^{-\eta \Delta t}\right)\left[\bar{\alpha} + \frac{(r-\rho)}{\eta}\right] + \sigma_2 \sqrt{\frac{1 - e^{-2\eta \Delta t}}{2\eta}} \varepsilon \sqrt{\Delta t}.$

Parameters	Notation
Initial patent value	V_0
Initial expected rate of growth for patent value	α
Initial volatility of patent value	σ
Half decay rate of the growth of the drift	η
Long-term drift rate of patent value	$\bar{\alpha}$
Time interval	Δt
Long-term patent value	$\bar{\mu}$
Duration	T
Patent value	V

6.2 Viewing a Generic Case

Here, the model is applied to the case of a pharmaceutical company negotiating a phase 2 patent license. Analysing the uncertainty in the life cycle of the patent's value in this case reveals the following uncertainties: (i) Although the potential sales of the patent are considered stable, the sales parameter is in fact a pinpoint estimate, and actual sales fluctuate over time. (ii) The duration of phase 3 is unknown. (iii) The life cycle must be considered to reflect the real-world setting.

If the company uses the NPV method to evaluate its patent, the effect of uncertainty cannot be considered because of the pinpoint parameters. Second, the NPV method assumes that revenue flows are pinpoint estimates and constant over time, which is unrealistic for the patent life curve. Given the background, the proposed model describes mean-reverting motions with uncertainties, and the S-shaped life cycle can be used.

Because companies treat patent negotiations as business secrets, obtaining actual case figures is difficult. Nevertheless, the proposed model can be applied easily by inputting different case settings. The model was applied to this case using data reported in the literature.

The starting value (V_o) of the patent in the initial R&D stage is set at $1 million, which indicates that, although the patent is promising, licensing is very risky in this stage. The sales growth (α) is used as a proxy for the patent value, that is, the drift rate is 10%. The patent value volatility (σ) is 8% annually, which reflects the uncertainty about annual sales, and the reversion rate (η) is set at 2% in this analysis. The long-term patent value (μ) can be derived from government or institutional surveys.

For example, if the population of patients requiring drug treatment is 1 million worldwide, the population can be indicated in terms of sales. Therefore, an equilibrium level of $50 million annual sales revenues is assumed in the stable stage of the patent life cycle. By managing forecasts after acquiring the phase 2 patent, the company can launch the new treatment 2–3 years after the manufacturing plants have been constructed and the process development has been completed. Within 5–7 years after the launch, the new treatment will grow exponentially and reach stable market sales. In this analysis, it is assumed that the duration (T) of the patent's life cycle is 20 years.

6.3 Interferon Beta 1a: A Real-World Case

Interferon beta-1a is a cytokine in the interferon family used to treat multiple sclerosis (MS). Avonex was approved in the US in 1996, and in the European Union in 1997, and is registered in more than 80 countries worldwide. It is the leading MS therapy in the US, with around 40% of the overall market, and in the EU, with around 30% of the overall market. It is produced by the Biogen-IDEC and has been marketed under the trade names 'Avonex' (Biogen) and 'Rebif' (Merck KGaA). Peak global sales have been around USD 5 bn (Avonex: 3 bn, Rebif: 2 bn) in the period 2013–15.

An analysis of interferon beta-1a/Avonex, based on the potential market and the price that Biogen was expected to charge, yielded a present value of USD 3.4 bn, prior to consideration of the initial development cost. The initial cost of developing the drug for commercial use was estimated at USD 2.875 bn.

At the time of this particular analysis, the duration of patent protection on Avonex was another 17 years, and the then current long-term treasury bond rate was 6.7%. Using an aggregated stock market analysis, the average variance in firm value for publicly traded biotechnology firms ('volatility') was found to be 0.224.

To stochastically estimate the patent value, the Black-Scholes formula adjusted for dividends has been used (as detailed in Sect. 6.1):

$$d_1 = \frac{\ln\frac{3422}{2875} + \left(0.0675 - 0.589 + \frac{0.224}{2}\right)\cdot 17}{9.4377\cdot\sqrt{17}}$$

$$N(d_1) = 0.872$$

$$d_2 = \frac{\ln\frac{3422}{2875} + \left(0.0675 - 0.589 - \frac{0.224}{2}\right)0.17}{9.4377\cdot\sqrt{17}}$$

$$N(d_2) = 0.2076.$$

The patent value is $C = 3422e^{-0.0579\times 17} \times 0.872 - 2875 \times e^{-0.067\times 17} \times 0.2076 = 907$ (USD mn).

Contrast this result with the net present value of this project:

$$NPV = 3422 - 2875 = 547 \text{ (USD mn)}.$$

Although the NPV of the patent yields only USD 547 mn, the Black-Scholes model evaluates the patent fat USD 907 mn. The higher value in the latter case means that the patent holder may take advantage in delaying launch and waiting for better market conditions. Less time to the end of patent life will decline its value because it will increase the cost of delay. As can be seen from this example, patent valuation using real options has led to a higher value than by using NPV. The effect would be even more marked if the NPV is near zero or negative. Hence, real option pricing models can be better value metrics than traditional methods in determining the value of intangible assets based on the benefits of bringing the asset owner.

7 The Future of Value and Valuation in Pharma

Category and capability leadership hold the keys to superior value creation and even survival in pharma. Companies that stick to the old model of diversifying assets and spreading R&D bets across many categories will likely find themselves running conglomerates of sub-scale businesses. As the innovation bar for attractive reimbursement rises, they will face low profitability and negative returns on R&D. Category leaders will have more resources to invest in product development, commercialization and acquisitions. Because assets owned by sub-scale companies will be worth more in the portfolios of market leaders, current owners will risk being consolidated by the winners. Copying today's proven business models does not guarantee future success. Inevitably, today's leaders will use their market influence to raise the bar for competitors. However, there is good news for companies still building their category leadership positions.

(1) Data shows that winning in pharma depends on scale within categories rather than across the broader pharma market. In an increasingly fragmented industry,

categories are often defined far more narrowly than the traditional therapeutic and disease areas. Over the past decade, for example, Astellas has achieved leadership positions in urology and transplants and is currently shaping a narrower category, uro-oncology. In the future, there will be many similar opportunities to define and lead new categories in pharma.
(2) It is seen that today's pharma category leaders only use a small fraction of the tools and tactics successfully employed in other industries. For example, the standard commercial model in pharma relies on unit-based pricing, a narrow product definition (pill or vial) and long-established promotional techniques. All three elements are ripe for disruption.
(3) Pharma companies still operate in a high-margin environment. As a result, they often focus on defending their positions rather than doing things differently.

Current leaders face a particular dilemma: leaders that change too early risk losing attractive cash flows from established business models; those that move too late risk being disrupted by emerging competitors. In the recent history of the industry, it seems that leaders have more often erred on the side of holding on to old models for too long, leaving room for more aggressive players to disrupt them. New and innovative business models across verticals can generate greater value and deliver better care for individuals. New and innovative business models are beginning to show promise in delivering better care and generating higher returns. The existence of these models and their initial successes are reflective of what we have observed in the market in recent years: leading organizations in the healthcare industry are not content to simply play in attractive segments and markets, but instead are proactively and fundamentally reshaping how the industry operates and how care is delivered. While the recipe across verticals varies, common among these new business models are greater alignment of incentives typically involving risk bearing, better integration of care, and use of data and advanced analytics.

The pharma industry continues to evolve, with potential disruptions affecting all parts of the value chain, from R&D to patient care. The future success of today's market leaders will be determined by how they react to these changes. Pfizer has already started to apply its commercial optimizer model in specialty businesses. And many companies struggle to repeat breakthrough innovation in a particular disease area, because competitors soon close the gap with similar products. To stay ahead of the competition, breakthrough innovators often evolve into disease solutions providers in the categories they helped create. In oncology, for example, Roche has been building a sophisticated business system on the strength of its breakthrough cancer therapies. Future winners will actively disrupt current business models, including their own. For example, pricing models will increasingly shift from per-pill pricing to outcome-based and at-risk models. Disease solution providers will move to own 'episodes of care', including diagnostics, drugs, devices and treatment protocols.

References

Ahlstrom, D. (2010). Innovation and Growth: How Business Contributes to Society (2010). Report, The Chinese University of Hong Kong (CUHK), available at SSRN. https://ssrn.com/abstract=2643390.

Banerjee, A., Bakshi, R., & Kumar, M. (2019). *Valuation of Patents?: A Classification of Methodologies*. Department of Human Resource Management, IIEST, Shibpur/Kolkata, India: Report.

Baudry, M., & Dumont, B. (2006). Patent renewals as options: improving the mechanism for weeding out lousy patents. *Review of Industrial Organization, 28*, 41–62. https://doi.org/10.1007/s11151-006-0001-0.

Behnke, N., Retterath, M., Sangster, T., Singh, A. (2014). *New Paths to Value Creation in Pharma*. Report, Bain & Co., Boston, MA

Buldyrev, S. V., Pammolli, F., Riccaboni, M., & Stanley, H. E. (2020). *The Rise and Fall of Business Firms: A Stochastic Framework on Innovation*. Creative Destruction and Growth (p. 9781107175488). Cambridge University Press.

Clark, E., Singhal, S., Weber, K. (2021). *The Future of Healthcare: Value Creation Through Next-Generation Business Models*. Report, McKinsey & Co., Chicago, IL

Deng, Y. (2011). A dynamic stochastic analysis of international patent application and renewal processes. *International Journal Industrial Organization, 29*(6), 766–777.

Dias, M.A. (2001). Selection of alternatives of investment in information for oil-field development using evolutionary real options approach. In *Proceedings of 5th Annual International Conference on Real Options*

Dixit, A.K., & Pindyck, R.S. (1994). *Investment under uncertainty*. Princeton University Press.

Goldenberg, D. H., & Linton, J. D. (2012). The patent paradox—new insights through decision support using compound options. *Technological Forecasting and Social Change, 79*, 180–185.

Kellogg, D., & Charnes, J. M. (2000). Real-Ootions valuation for a biotechnology company. *Financial Analysts Journal, 56*(3), 76–84.

Leone, M. I., & Orianim, R. (2021). *The Option Value Of Patent Licenses*. Retrieved April 17, 2021 from www.epip.eu/theoptionvalueofpatentlicenses.

Meeks, M. T., & Eldering, C. A. (2010). Patent valuation: aren't we forgetting something? *Making the Case for Claims Analysis in Patent Valuation by Proposing a Patent Valuation Method and a Patent-Specific Discount Rate Using the CAPM, Northwestern Journal of Technology and Intellectual Property, 09*(03), 3212.

Myers, S. C., & Howe, C. D. (1997). *A life-cycle financial model of pharmaceutical R & D, Working Paper 4197*. Sloan School of Management, Massachusetts Institute of Technology. Cambridge, MA: Program on the Pharmaceutical Industry.

Reitzig, M. (2000). *Methods For Patent Portfolio Valuations*. Retrieved January 22, 2022 from www.oecd.org/sti/sci-tech/35428864.pdf.

Schwartz, E. S., & Moon, M. (2000). Rational pricing of Internet companies. *Financial Analysts Journal, 56*, 62–75.

Schwartz, E. S. (2004). Patents and R&D as real options. *Economic Notes, 23*, 24.

Sebastian, H., Legler, E., & Lichtenthaler, U. (2010). Determinants of patent value: Insights from a simulation analysis. *Technological Forecasting and Social Change, 77*, 01–19.

Sereno, L. (2010). *Real Options Valuation of Pharmaceutical Patents. A Case Study*. Retrieved December 22, 2012 from SSRN 1547185.

Sohn, S. Y., Lee, W. S., & Ju, Y. H. (2013). Valuing academic patents and intellectual properties: Different perspectives of willingness to pay and sell. *Technovation, 33*, 13–24.

Triest, S., & Vis, W. (2007). Valuing patents on cost-reducing technology: A case study. *Production Economics, 105*, 282–292.

Villiger, R., & Bogdan, B. (2005). Getting real about valuations in biotech. *Nature Biotechnoloy, 23*(4), 423–8.

Wu, L., & Wu, L. (2011). Pharmaceutical patent evaluation and licensing using a stochastic model and Monte Carlo simulations. *Nature Biotechnology, 29*(9), 798–801.

Michael Blankenagel Senior Lecturer at Lucerne University of Applied Sciences and Arts. He specializes in Corporate Performance Management and Research Design. Beforehand, he worked among other things as CFO of an international corporation, Managing Director and Management Consultant.

Jung Kyu Canci Senior lecturer and researcher at University of Basel and of Applied Science in Lucerne. His research is in pure mathematics, Number Theory with particular interests in Arithmetic of Dynamical Systems, and in applied mathematics, Stochastic Processes in Finance. He is also the founder of several companies.

Philipp Mekler Trained in both biochemistry (Ph.D.) and mathematics (MSc), with more than forty years of experience in the Pharma/Life Science sector in R&D, sales/marketing, business analysis and bio-business finance/venture capital (in CH, US, & IL). Currently at Roche International as Strategy Advisor for a Data & Analytics unit.

Open Access This chapter is licensed under the terms of the Creative Commons Attribution 4.0 International License (http://creativecommons.org/licenses/by/4.0/), which permits use, sharing, adaptation, distribution and reproduction in any medium or format, as long as you give appropriate credit to the original author(s) and the source, provide a link to the Creative Commons license and indicate if changes were made.

The images or other third party material in this chapter are included in the chapter's Creative Commons license, unless indicated otherwise in a credit line to the material. If material is not included in the chapter's Creative Commons license and your intended use is not permitted by statutory regulation or exceeds the permitted use, you will need to obtain permission directly from the copyright holder.

Limited Commercial Licensing Strategies: A Piecewise Deterministic Differential Game

Domenico De Giovanni and Jung Kyu Canci

1 Limited Commercial Licenses in the Pharmaceutical Industry

In 2021 the members of the Word Trade Organization signed the so-called TRIPS (Trade-Related Aspects of Intellectual Property Rights). https://www.wto.org/english/docs_e/legal_e/27-trips_01_e.htm.

This declaration gives the opportunity to governments of developing and least developed countries to issue limited commercial licenses (denoted by LCL) also know in literature with the name *compulsory licenses* (CL. According to its name, a limited commercial license is a government authorized non-voluntary license from a patent holder (henceforth H) to a third party, usually a generic producer (denoted by G). The government of a country C, which in what follows will be always a developing or least developed one, may issue a CL for a drug, covered by intellectual property, only under certain conditions. The most relevant one is when the sales S_H of the patent-holder H of a drug D does not reach an expected target sale denoted by $\tilde{S}$, in the situation where H is the unique producer of the drug H in the country. Therefore, a CL is issued if in a regime of monopoly a certain drug as a selling result, that is below the expected one (i.e., $\tilde{S}$). Often the sale target $\tilde{S}$ is not public, so it is considered a random variable and it is set taking into consideration, for example, the expected case of cases of the disease, for which the drug was developed.

D. De Giovanni
University of Calabria, Calabria, Italy
e-mail: ddegiovanni@unical.it

J. K. Canci (✉)
Lucerne University of Applied Sciences and Arts, Lucerne, Switzerland
e-mail: jungkyu.canci@hslu.ch

© The Author(s) 2023
J. K. Canci et al. (eds.), *Quantitative Models in Life Science Business*,
SpringerBriefs in Economics, https://doi.org/10.1007/978-3-031-11814-2_2

In the article Sarmah et al. (2020) the authors considered different scenarios, where a pharma manufacturer invests in R&D, obtaining the intellectual property of a new drug. This patent can be subjected to a limited commercial license. The author considered four different periods:

- t_0 In this period the pharma company H invests in R& D for developing a new drug. At the end of this period H owns the patent for the new drug.
- t_1 In this period the company H launched the drug in the market of a country C. During this period the market is a monopoly, because H is the unique company having the right to produce the drug.
- t_2 In this period the government of the country C may issue a limited commercial license. This happens in the case the selling result S_{H_1} does not reach the selling target $\tilde{S}$ in the previous phase t_1. If the CL is issued, a competitor G (generic producer) also has the right to produce the drug, H must license G to produce the drug, but G has to pay a royalty fee to H. If in period t_1 the selling target $\tilde{S}$ is reached, then the CL is not issued and so H operates in a monopoly system.
- t_3 The intellectual property of H expires and H must operate in a free market, where a generic producer may produce the drug.

Compulsory licensing has been the subject of interest in numerous recent studies. For example, Scherer (1977) discusses how regulators can use limited commercial licensing to restore competition in industries. Aoki and Small (2004) views compulsory licensing as a tool of anti-competitive practices and shows that it creates significant losses. Seifert (2015) documents that compulsory licensing decreases incentives for innovations, but creates benefits to consumers and total welfare. Bertran and Turner (2017) suggest that a, in duopoly, suitable royalty payments are required to improve social welfare. Bond and Saggi (2014) analyze the role of compulsory licensing in determining consumer access to a patented product sold by a patent-holder. They suggest that compulsory licensing guarantees consumer access to the patented product and increment the chances of voluntary licensing and results in the patent-holder switching from voluntary licensing to entry. In the same spirit, Stavropoulou and Valletti (2015) find that the overall welfare effects of compulsory licensing are positive even if taking into account innovation incentive. Finally, empirical studies analyze the effects in terms of incentives to innovate compulsory licensing. For instance, Baten et al. (2017); Moser and Voena (2012) suggest that compulsory licensing pushes innovators to create new patents.

In the article Sarmah et al. (2020), the authors analyze a dynamic game between the patent-holder H and a generic producer G. They considered three different scenarios, described in the picture below

Scenario B (benchmark): The CL is not issued

t_0	t_1	t_2	t_3
Firm H invests in R&D knowing that no CL will be issued	Firm H sets the price, p_1	-Firm H sets the price, p_2 -Firm G does not have the license to produce the generic drug	-Firm H sets the price, p_3 -Firm G sets the price, w_3

Scenario D (deterministic): The CL is issued

t_0	t_1	t_2	t_3
Firm H invests in R&D knowing that the CL will be issued in t_2	Firm H sets the price, p_1	-Firm H sets the price, p_2 -Firm G receives the license to produce the generic drug, sets the price w_2 and transfers royalty ϕ	-Firm H sets the price, p_3 -Firm G sets the price, w_3

Scenario S (stochastic): The information about issuing the CL is revealed at time t_2

t_0	t_1	t_2	t_3
Firm H invests in R&D. There is uncertainty on whether a CL will be issued. The likelihood of the issue depends on H's sales in t_1	Firm H sets the price, p_1, still under uncertainty	Information is revealed	-Firm H sets the price, p_3 -Firm G sets the price, w_3

The authors in Sarmah et al. (2020) compare the different three above scenarios. The incertitude due to the risk of a compulsory license makes an investment in R&D less attractive. The issue of a compulsory license should improve the access to a new drug, but the economic-market setting should appropriately remunerate pharma companies and guarantee the profitability of the investments in R&D.

The authors in Sarmah et al. (2020) have shown that a sufficient high royalty, paid in period t_2 in the case of a CL, could at the same guarantee enough good condition to H, the intellectual-property holder, and assure enough sustainable access to the drug.

The outline of this paper is as follows. In the next Sect. 1.1 we study the problem of determining, whether the CL will be issued. In Sect. 2 we model concerning a CL by using Differential Game Theory, as presented in Dockner et al. (2000). In Sect. 3 we explain how the model can be solved. Section 4 gives some perspective for future work on the subject.

1.1 Probability of Issuing a CL

In the stochastic scenario it is extremely important to predict whether the limited commercial license will be issued or not. We consider the cumulative function $F_{\tilde{S}}(x) = \text{Prob}(x \leq \tilde{S})$.

By denoting S_{H_1} the selling result in phase t_1, than the probability that the CL is issued is equal to $F_{\tilde{S}}(S_{H_1}) = \text{Prob}(S_{H_1} \leq \tilde{S})$, because the CL is issued in the case the sale target result $\tilde{S}$ is not reached in phase t_1.

Therefore, we need to estimate which value can have $\tilde{S}$, and whether the selling result S_{H_1} may reach the value $\tilde{S}$.

In Sect. 5 in Sarmah et al. (2020), the authors model H's belief about government's target by using a log-logistic random variable, thus they assume

$$F_{\tilde{S}}(x) = \frac{1}{1+\left(\frac{x}{a}\right)^{-b}}$$

for suitable values of a and b depending on the market setting, in their article the authors have set $a = b = 2$.

Therefore, we can estimate $\tilde{S}$ by calculating the expected value of the log-logistic random variable.

A new idea to estimate, whether the sailing result reaches the value of the estimated value for $\tilde{S}$ is given by using a counting process on the random variable S_{H_1}.

One of the easiest models of the counting process is the so-called (stationary) Poisson process. We consider a subdivision of an interval in sub-intervals with endpoints $a_1 < a_2 < \ldots < a_k$. For any index $1 \leq i \leq k-1$ we consider a time $b_i \in (a_i, a_{i+1})$ and an integer n_i. We denote by $N(a_i, b_i]$ the number of events of the process happening in the interval $(a_i, b_i]$. In the Poisson process, we have

$$\text{Prob}(N(a_i, b_i] = n_i, i = 1, \ldots, k-1) = \prod_{i=1}^{k-1} \frac{(\lambda \cdot (b_i - a_i))^{n_i}}{n_i!} e^{-\lambda \cdot (b_i - a_i)}, \quad (1)$$

where λ is the parameter, which characterizes the Poisson process.

In the model, the time-points a_i's and b_i's can represent some marketing actions, workshops, scientific-information meetings, etc.

Note that the above formula (1) appears slightly simpler in the case $b_i = a_{i+1}$ for all index i. The model is called stationary because the distributions are stationary, in the sense they do not depend on the numbers a_i's and b_i's, but on the differences $b_i - a_i$.

Therefore, we can subdivide the period t_1 in the CL-model above into several sub-intervals and estimate the corresponding selling sub-results n_i in each sub-period and with the above formula to calculate how is realizable (i.e., probable) the minimal goal $\tilde{S}$.

As a first approximation we can consider a unique period t_1 (with no subdivision), obtaining as a model a classical (univariate) Poisson distribution.

Another possible way is to consider the notion of the Hawkes process. The idea is that the selling of a new drug in a country is a self-exciting counting process. Indeed, by assuming the trivial hypothesis that the drug has a positive effect in treating the corresponding disease (for which the drug was developed), then the selling of a drug dose positively influences the selling of the next doses.

2 A Differential Game of Limited Commercial Licensing

In an environment in which a limited commercial license might be issued at some future point in time, the problem of drug pricing and R&D in product innovation depends on the stage of the process of issuing a CL. In each different stage of this process, the actors involved are allowed to take different actions, according to which stage is currently active. To model the different behavior of the actors involved in each stage of the limited commercial license's life, we make use of the framework of piecewise-deterministic differential games described in Dockner et al. (2000). In this setup, a discrete set of *modes*, M, models have different stages of the system. In each mode, players are allowed to take some actions. Switches between modes are randomly driven by a continuous-time Markov chain with values in M. The probability of switching between two modes of the systems in general depends on the actions taken by the players and the state variables of the system. While we refer to the above-mentioned book of Dockner et al. (2000) for a detailed treatment of piecewise-deterministic differential games, in what follows we describe the details of our model of pricing and product innovation in the uncertain environment of a limited commercial license. In the following parts we will use the standard notation as the one used in the book of Dockner et al. (2000) (Chapter 8).

2.1 Preliminaries

We consider a piecewise-deterministic game with two players. The first, which we will refer to as the innovator (I), who has developed a patent for a new drug and started selling the product in a new, underdeveloped, country. The second, which we will call the generic producer (G), we want to exploit the patent of the innovator. On top of both players, there is a regulatory authority (the government or similar) in charge of evaluating the needs of the country in terms of the new drug, and eventually issuing a CL. We will assume that the issue or not of a CL depends only on sales of the new drug, leaving aside political/economical reasons.

We now introduce a set $M = \{1, 2, 3\}$ of modes of the system. Broadly speaking, M identifies the stages involved during the life of a new drug with the following (this of course can be changed according on how much emphasis we want to put in each stage of development):

- 1 indicates the stage in which the innovator launches a new drug on the market;
- 2 models the situation in which a CL is currently active;
- 3 represents the end of the process, where the patent expires.

Next, we introduce the following notation ($i \in \{I, G\}$): Let $x_i = x_i(t)$ represent the cumulative sales of player i up to time t and $p_i = p_i(t)$ be the price of the drug decided by player i at time t. Moreover, let $A_I = A_I(t)$ model the innovator's effort in product innovation at time t. The evolution of the cumulative sales of each player is governed by the so-called players' instantaneous sales functions, f_i, that is

$$\dot{x}_i(t) = \frac{dx_i(t)}{dt} = f_i(x_I, x_G, p_I, p_G, A_I, m) \qquad (2)$$

for $m \in M$. We observe that the functions that govern instantaneous sales depend on the mode of the system. This is a feature of piecewise-deterministic games that allows to take into account the different phases of the development of the limited commercial license.

2.2 Profits and Sales in Each Regime

We now specify how sales evolve and what profits the players obtain in each mode of the system:

- In mode 1, since no limited commercial license exists, the generic producer is not allowed to sell the drug, and $f_G(x_I, x_G, p_I, p_G, A_I, 1) = 0$. On the other hand, the innovator operates in a situation of monopoly, as they are the only seller in the market. We assume that their instantaneous sales are

$$f_{I_1} = f_I(x_I, x_G, p_I, p_G, A_I, 1) = a - b_I p_I + \theta A_I, \qquad (3)$$

where a is the market potential, b_I is the market elasticity with respect to the price of the patented product, and θ is the market sensitivity to improvements in the product. In this regime, the innovator's instantaneous profit is modeled as the difference between instantaneous market revenues and costs for production and product development, that is

$$\pi_I(p_I, A_I, 1) = (p_I - c_I) f_{I,1} - l A_I^2$$

being c_I the marginal production cost for production and l the coefficient for the (quadratic) cost in product development.

- In mode 2, the limited commercial license has been issued. This means that the generic producer has entered the market in exchange for a royalty to be paid to the innovator. The sales functions are described as follows:

$$f_{G_2} = f_G(x_I, x_G, p_I, p_G, A_I, 2) = a - b_G p_G + \theta A_I \qquad (4)$$
$$f_{I_2} = f_I(x_I, x_G, p_I, p_G, A_I, 2) = a - b_I p_I + \theta A_I \qquad (5)$$

being p_G, b_G the price assigned by the generic producer and their price elasticity, respectively. The profit functions take into account the royalty that G needs to pay to the innovator, as described in the following

$$\pi_G(p_G, 2) = (p_G - c_G - R) f_{G_2} \qquad (6)$$
$$\pi_I(p_I, A_I, 2) = (p_I - c_I) f_{I,2} - l A_I^2 + R f_{G_2}, \qquad (7)$$

where R is the royalty that G needs to pay for a unit of the drug sold.
- In the last stage of the game, as the patent has expired, the generic producer enters the market without the need for a limited commercial license. Hence, a classical price competition takes place. Sales functions and profits are described, respectively, as

$$f_{G_3} = f_G(x_I, x_G, p_I, p_G, A_I, 3) = a - b_G p_G + \theta A_I \qquad (8)$$
$$f_{I_3} = f_I(x_I, x_G, p_I, p_G, A_I, 3) = a - b_I p_I + \theta A_I, \qquad (9)$$

and

$$\pi_G(p_G, 3) = (p_G - c_G)f_{G_3} \quad (10)$$
$$\pi_I(p_I, 3) = (p_I - c_I)f_{I_3}. \quad (11)$$

Here, we are implicitly assuming that as the patent expires the innovator stops research in product innovation.

2.3 Switching Between Stages

In the setup we have in mind, switches between one mode of the system to another are endogenous to the problem itself. In particular, when the innovator enters a market with a new drug, they does not know if and when a limited commercial license would be issued at some future point in time. For this reason, we now introduce a continuous-time Markov chain with values in M that drives switches between modes, as it follows. Let us fix a probability space $(\Omega, \mathbb{F}, \mathscr{P})$ and define a continuous-time Markov chain $\xi = \xi(t) : \Omega \times [0, \infty) \to M$, which generates the information flow represented by the family of σ-algebras $\{\mathscr{F}_t\}_{t \geq 0}$. This stochastic process is deterministic everywhere but at random times $\tau_i, i = 1, 2, \ldots$ where ξ jumps. The instant of times in which the chain jumps model the switches between regimes. The randomness of this process mimics the uncertainty of the market players about the eventual issue (and the time of issue) of the limited commercial license.

The probability law which drives the jumps of ξ, and hence the switches between the stages of developments of the limited commercial license, is described by a set of function $\lambda_{m,n} : \Omega \to \mathbb{R}_+$, for $m, n \in M, m \neq n$, defined as

$$\lambda_{m,n}(t, x_I, X_G, p_I, P_G, A_I) = \lim_{dt \to 0} \frac{\mathbb{P}\left(\xi(t + dt) = n | \xi(t) = m\right)}{dt} \quad (12)$$

which defines the conditional probability, measured at time t, of switching from mode m to mode n of the system in an infinitesimal amount of time dt to be proportional to dt. Such functions, usually known as hazard functions, are part of our modeling framework, as they completely characterize the Markov chain itself.

In our setup, the system might go from mode 1 (the innovator has entered the market) to mode 2 (the limited commercial license has been issued) if the sales of drug are sufficiently low. It is thus meaningful to assume that the corresponding hazard function $\lambda_{1,2}$, which roughly speaking defines the probability of issuing the CL, is a decreasing function of the cumulative sales in mode 1 of the system, as follows:

$$\lambda_{1,2} = \lambda_{1,2}(t, x_I, X_G, p_I, P_G, A_I) = \frac{\delta_{1,2}}{x_I}.$$

On the other hand, the CL may never be issued if the drug reaches a considerable amount of potential patients. In such cases, the system might go directly from stage 1 to stage 3, where the patent has expired. In the same spirit of our reasoning above, the hazard function for such kinds of switches should be directly proportional to the amount of sales in mode 1, that is:

$$\lambda_{1,3} = \lambda_{1,3}(t, x_I, X_G, p_I, P_G, A_I) = \delta_{1,3} x_I.$$

To close our model, we need to define the intensity of switches between mode 2 and mode 3 of the system. Since the expiration of the patent is independent on the sales of the product, we just assume a constant hazard function, that is

$$\lambda_{2,3} = \lambda_{2,3}(t, x_I, X_G, p_I, P_G, A_I) = \delta_{2,3}.$$

2.4 The Problem

The problem is defined as a differential game, in which both players choose their strategies so as to maximize the discounted expected value of future profits. Let r be the rate used by both players to discount their profits. Define the objective functionals of both players, in each mode of the system m, as follows:

$$J_G(x, y, p_I, p_G, A_I, m) = \mathbb{E}\left(\int_0^\infty e^{-rs} \pi_G(p_G, \xi(s)) | x_I(0) = x; x_G(0) = y; \xi(0) = 1\right)$$

$$J_I(x, y, p_I, p_G, A_I, m) = \mathbb{E}\left(\int_0^\infty e^{-rs} \pi_I(p_I, A_I, \xi(s)) | x_I(0) = x; x_G(0) = y; \xi(0) = 1\right)$$

Then, both players choose their pricing and product development strategies so as to maximize the objective functional above under the dynamic constraints given by (12) and

$$\dot{x}_i(t) = \frac{dx_i(t)}{dt} = f_i(x_I, x_G, p_I, p_G, A_I, m) \quad i \in I, G; m \in M. \quad (13)$$

3 Solving the Model

To solve the problem, one must first choose the type of strategies available to the players. As it is standard in the theory of stochastic (in this case piecewise deterministic) differential games, two classes of strategies are available. Open-loop strategies correspond to an entire path chosen for the control variable, while with closed-loop strategies (also known as feedback or Markov strategies) the control variable are a function of the observed level of the state space. The latter class of strategies is more appealing as it appears to be more realistic than open-loop strategies. On the other hand, solving for feedback strategies is, in general, much more difficult. In the case of the problem at hand, we do believe that it can be solved in closed-loop strategies, with the help of the paradigm of backward induction and numerical techniques.

Let us begin with feedback strategies, also known as Markovian strategies. In this setup, the optimal controls are expressed in terms of the state variables, that is $p_i(t) = \psi_i(x_I(t), x_G(t))$. This means that at each time players perform an action based on the observed state of the system. In this case, the solution of the problem can be expressed in terms of a system of Hamilton-Jacobi-Bellman equations. Let $V_i(x, m)$ denote the value function of player $i \in \{I, G\}$ in regime $m \in \{1, 2, 3\}$, with $x = (x_I, x_G)$. Then, for $V_i(x, m)$ to be the solution of the piecewise-deterministic differential game, they must solve the following system:

$$rV_i(x, m) = \max_{p_i} \left\{ \pi_i(p_i, m) + \frac{d}{dx_I} V_I(x, m) f_I(x, p_I, p_G, h) + \frac{d}{dx_G} V_G(x, m) f_G(x, p_I, p_G, h) + \sum_{n \neq m} \lambda_{n,m} (V_i(x, n) - V_i(x, m)) \right\} \quad (14)$$

Let us now focus on open-loop strategies. In this case, the actions of the players are expressed in terms of time, the regime of the system and the state of the system at the last observed switch, that is $p_i(t) = \phi_i(m(s(t)), x(s(t)), t - s(t))$, where $s(t)$ denotes the time in which the last switching of regime occurred before time t. In such case, the value functions depend explicitly on time. Thus, for $V_i(x, m, t)$ to be the solution to the problem under piecewise open-loop strategies, they must solve the following system of Hamilton-Jacobi-Bellman equation:

$$rV_i(x, m, t) - \frac{d}{d_t} V_i(x, m, t) = \max_{p_i} \left\{ \pi_i(p_i, m) + \frac{d}{dx_I} V_I(x, m, t) f_I(x, p_I, p_G, h) + \frac{d}{dx_G} V_G(x, m, t) f_G(x, p_I, p_G, h) + \sum_{n \neq m} \lambda_{n,m} (V_i(x, n, 0) - V_i(x, m, t)) \right\} \quad (15)$$

The systems of equation in (14) and (15) describe sufficient conditions for the solution of the problem sketched in Sect. 2. However, both conditions do not admit closed-form solution, so that we must use a mix of analytical and numerical techniques to derive the optimal policies followed by the players. One approach is to discretize the system of Hamilton-Jacobi-Bellman equation by means of a semi-lagrangian approach (Falcone and Ferretti 2013).[1] This entails splitting the time horizon into a sequence of equidistant steps. We then approximate the variable $x(t)$ by means of the sequence x_n^h. We will make use of the conditional probability that the process will jump from mode i to mode j in an analogous time step, which we approximate as

---

[1] Applications of Semi-Lagrangian schemes in economics and management can be found in Santos and Vigo-Aguiar (1998), Grüne and Semmler (2004), De Giovanni and Lamantia (2018).

$$P^h_{x,i,j}(q) = 1 - e^{-h\lambda_{i,j}(x,q)}. \tag{16}$$

The continuous-time optimal control problem is thus replaced by the following first-order discrete-time approximation

$$V^h(x,i) = \max_{q_1,q_2,\ldots} E_{x,i} \left\{ \sum_{l=0}^{\infty} \sum_{n=N_l}^{N_{l+1}-1} h\beta^n \pi^{\xi_l}(x_n^h, q_{n-N_l}) \right\}, \quad i \in I, \tag{17}$$

where we set the discount factor $\beta = e^{-\omega h}$. Camilli (1997) shows that $V^h(x,i)$ satisfies the following dynamic programming equation

$$V^h(x,i) = \max_q E_{x,i} \left\{ h\pi^i(x,q) + \beta V^h(x_1^h, i) \right\}. \tag{18}$$

Finally, (18) gives the discrete-time infinite dimensional system of equations satisfied by the value functions $V^h(x) \stackrel{def}{=} \{V^h(x,i) : i \in I\}$

$$V^h(x,i) = \mathcal{N}_i(V^h(x)) \quad i \in I, \tag{19}$$

where the dynamic programming operators $\mathcal{N}_i(\cdot)$ are defined by

$$\mathcal{N}_i(V^h(x)) \stackrel{def}{=} \max_q \{h\pi^i(x,q) + \beta P^h_{x,i}(q)V^h(x + hG(x,q,i),i) + \sum_{j \neq i} P^h_{x,i,j}(q)V^h(x,j).\} \tag{20}$$

Problem (20) is still infinite dimensional in the state variable. However, we can convert it into a set of finite-dimensional equations by partitioning the state space into a grid $\Gamma = \{x_k : k = 1, \ldots, K\}$ and solve (20) only for $x \in \Gamma$. To make the scheme operative, we need to reconstruct the values $V^h(x_k + hf(x_k, \alpha, i), i)$ since in general the points $x_k + hf(x_k, \alpha, i)$ do not coincide with any point of Γ.

4 Perspective

We have outlined a novel framework for the analysis of investments of a large company under the risk of limited commercial licensing. The framework captures all the key features on the subject and allows researchers to tackle important research questions such as

– How does the risk-limited commercial licensing affect the investment patterns of innovators in research and development?

- How does limited commercial licensing impact in the pricing of new, licensed, products?
- Can a regulator devise a licensing mechanism that boosts investments in research and developments and protect public welfare at the same time?

In future research, we plan to exploit the setup produced in this paper with the aim of tackling the research questions above.

References

Aoki, R., & Small, J. (2004). Compulsory licensing of technology and the essential facilities doctrine. *Information Economics and Policy*, 16113–16129.

Baten, J., Bianchi, N. Moser, & P. (2017). Compulsory licensing and innovation–historical evidence from German patents after WWI. *Journal of Development Economics*, 126231–126242.

Bertran, F. J L., & Turner, J L. (2017). Welfare-optimal patent royalties when imitation is costly. *Journal of Economic Behavior & Organization*, 137457–137475.

Bond, E. W., & Saggi, K. (2014). Compulsory licensing, price controls, and access to patented foreign products. *Journal of Development Economics*, 109217–109228.

Camilli, F. (1997). Approximation of integro–differential equations associated with piecewise deterministic process. *Optimal Control Applications and Methods*, 18423–18444.

De Giovanni, D., & Lamantia, F. (2018). Dynamic harvesting under imperfect catch control. *Journal of Optimization Theory and Applications*, 1761252–1761267.

Dockner, E. J., Jorgensen, S., Van Long, N., & Sorger, G. (2000). *Differential games in economics and management science*. Cambridge University Press.

Falcone, M. and Ferretti, R. (2013). Semi-lagrangian approximation schemes for linear and hamilton-jacobi equations. *SIAM*.

Grüne, L., & Semmler, W. (2004). Using dynamic programming with adaptive grid scheme for optimal control problems in economics. *Journal of Economic Dynamics and Control*, 282427–282456.

Moser, P., & Voena, A. (2012). Compulsory licensing: Evidence from the trading with the enemy act. *The American Economic Review*, 1021396–1021427.

Santos, M. S., & Vigo-Aguiar, J. (1998). Analysis of a numerical dynamic programming algorithm applied to economic models. *Econometrica*, 662409–662426.

Sarmah, A., De Giovanni, D., & De Giovanni, P. (2020). Compulsory licenses in the pharmaceutical industry: Pricing and r&d strategies. *European Journal of Operational Research*, 28231053–28231069. https://EconPapers.repec.org/RePEc:eee:ejores:v:282:y:2020:i:3:p:1053-1069.

Scherer, F. M. (1977). *The economic effects of compulsory patent licensing* (No. 2). New York University, Graduate School of Business Administration, Center for the Study of Financial Institutions.

Seifert, J. (2015). Welfare effects of compulsory licensing. *Journal of Regulatory Economics*, 483317–483350.

Stavropoulou, C., & Valletti, T. (2015). Compulsory licensing and access to drugs. *The European Journal of Health Economics*, 16183–16194.

Domenico De Giovanni Associate professor of Mathematical Finance at the Department of Economics, Statistics and Finance—University of Calabria. He is currently Deputy Director of the Master in Finance and Insurance. His research activities are: Finance and Insurance modeling, Dynamic Game theory, Optimal Control and Investment Theory.

Jung Kyu Canci Senior lecturer and researcher at University of Basel and of Applied Science in Lucerne. His research is in pure mathematics, Number Theory with particular interests in Arithmetic of Dynamical Systems, and in applied mathematics, Stochastic Processes in Finance. He is also the founder of several companies.

Open Access This chapter is licensed under the terms of the Creative Commons Attribution 4.0 International License (http://creativecommons.org/licenses/by/4.0/), which permits use, sharing, adaptation, distribution and reproduction in any medium or format, as long as you give appropriate credit to the original author(s) and the source, provide a link to the Creative Commons license and indicate if changes were made.

The images or other third party material in this chapter are included in the chapter's Creative Commons license, unless indicated otherwise in a credit line to the material. If material is not included in the chapter's Creative Commons license and your intended use is not permitted by statutory regulation or exceeds the permitted use, you will need to obtain permission directly from the copyright holder.

Partnership Models for R&D in the Pharmaceutical Industry

Gianpaolo Iazzolino and Rita Bozzo

1 Introduction

The Research and Development [R&D] process in the pharmaceutical industry is particularly complex and demanding. The pharmaceutical company's ability to be innovative and competitive is determined by the success of the R&D process, since the result of research coincides with the organization's source of profit. In addition to its intrinsically complex nature, the process is very sensitive to economic, technological, and social factors, which leads the pharmaceutical company to face continuous challenges. The world today seems to be moving much faster compared to a few decades ago, with disruptive innovations that must be responded to with a minimum reaction time in order to survive in the market. This requires companies of all sectors to adapt faster, to be flexible, and always ready to seize opportunities and defend themselves against unexpected and unpredictable threats that can emerge in the rapid course of events. In the past, the pharmaceutical sector found it difficult to adapt to these new conditions, remaining rooted in traditional systems and showing a certain reluctance toward the innovations that have totally changed many other industries. This has led to a decreasing trend in the productivity of R&D investments. As Greiner's theory of evolution shows, crises drive growth. Also in this case, the temporary crisis has given an input to the companies in the sector for a greater opening toward new business models, causing significant changes especially in the R&D structure. However, progress in this direction may not be sufficient to put the pharmaceutical industry in a completely safe position, as it yet has to be apt for new challenges.

G. Iazzolino (✉)
Department of Mechanical, Energy and Management Engineering, University of Calabria, Building 41/C, 87036 Rende, Cosenza, Italy
e-mail: gianpaolo.iazzolino@unical.it

R. Bozzo
University of Calabria, Rende, Cosenza, Italy
e-mail: rita.bozzo97@gmail.com

© The Author(s) 2023
J. K. Canci et al. (eds.), *Quantitative Models in Life Science Business*, SpringerBriefs in Economics, https://doi.org/10.1007/978-3-031-11814-2_3

2 The Problem of R&D Efficiency of Pharmaceutical Companies

2.1 Eroom's Law

R&D efficiency is typically measured as the ratio of output to input. The most commonly used outputs are the number of approved NMEs (New Molecular Entities), the number of scientific publications, patent applications, or patents granted (Schuhmacher et al. 2021). The costs of the R&D activity or the number of personnel employed in these activities are typically used as inputs. However, correctly measuring inputs and outputs of pharmaceutical research and development is quite difficult, given the complexity of the process, which includes multiple and heterogeneous sources of knowledge and lasts for several years. Furthermore, the process is more than ever influenced by external sources such as collaborations with public and private institutions, partnerships, and knowledge spillovers.

In recent decades, a sharp decline in R&D efficiency has taken place within the pharmaceutical industry as a result of the increasing complexity of R&D activities. Although investments in pharmaceutical R&D have seen a significant increase in recent years, the production of new approved drugs has instead slowed down. This caused the decline in efficiency that led to the creation of a new term, the "Eroom's law", in which the word "Eroom" is "Moore" read backward, to highlight the contrast with the advances of other forms of technology (like transistors) over time. The term was coined by researchers from Sanford Bernstein (UK), following the detection of an exponential increase in the overall cost of research and development on new drugs approved by the FDA over the past 60 years. They discovered that the number of new drugs approved by the US FDA per billion dollars of research and development spending in the pharmaceutical industry has halved approximately every 9 years since 1950 (Scannell et al. 2012) (Fig. 1).

The same researchers who observed this dynamic and who coined the expression "Eroom's law" also provided an explanation of the main causes of this decreasing trend. The first is called "better than the Beatles problem" referring to the fact that it would be difficult for new pop songs to be successful if they had to be necessarily better than the Beatles. What happens to new drugs is very similar: as the basket of approved and marketed drugs is already rich in effective medicines, it becomes increasingly difficult to develop new ones that are on par or better in terms of effectiveness. This discourages research and development in some areas that have already been explored and instead encourages the search for drugs that treat more difficult diseases and that hence face greater obstacles to approval and adoption. The second cause is the so-called "cautious regulator problem", which refers to the progressive reduction of risk tolerance by drug regulatory agencies. Whenever there is a negative event such as the removal of a drug from the market for safety reasons, the regulations tighten, making R&D more complex and time consuming, therefore more expensive. While in the past there was less risk aversion, today patient safety comes first, even if this leads to increased costs and a slowdown in innovation. A further cause is the

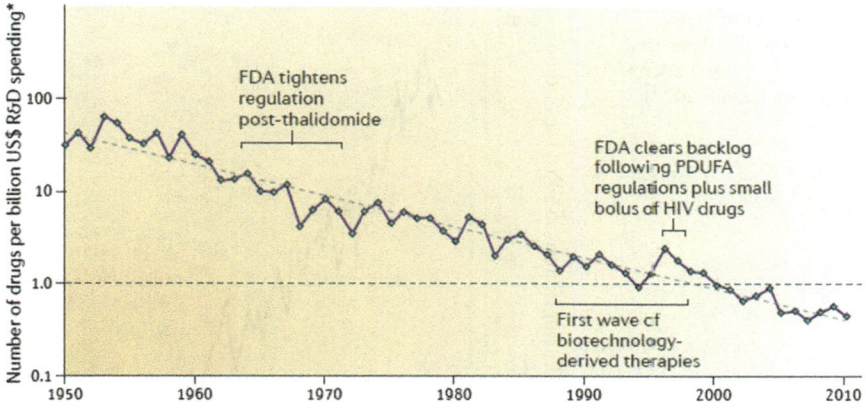

Fig. 1 R&D efficiency trend in the pharmaceutical industry. *Source* Scannell et al. (2012)

"throw money at it tendency", namely the tendency by pharmaceutical companies to increase R&D inputs, adding personnel, and investing additional resources with the illusory aim of improving company competitiveness. Today this trend seems to be decreasing, as many R&D costs are being cut, and productivity does not seem to suffer. Finally, the latest cause of Eroom's law is the "basic research-brute force bias", or the tendency to think that an improvement in basic research and in screening methods performed in the first steps of the standard discovery and pre-clinical research, can increase the safety and efficiency of clinical trials. However, this belief proved to be fallacious; in fact, the probability that a drug successfully passes clinical studies has remained almost constant for 50 years, and the overall efficiency of R&D activities has therefore decreased (Scannell et al. 2012).

Excessively long lead times mainly affect cost increases and the risk of industry rivalry (Schumacher et al. 2016). The cost increase is due to the cost capitalization, since these costs are incurred many years before the launch of any successful drug. Furthermore, excessively long timescales reduce the probability of being first on the market, since many companies focus on the same targets and compete only on time. The causes of the increasing R&D times are the long and strict regulation procedure to ensure the safety and efficacy of drugs and the interest of research increasingly oriented toward new and complex therapeutic areas aimed to differentiate from the competition. This implies a greater number of failures and therefore increasing costs. Another important consequence is the decrease in production in terms of new approved drugs. The majority of R&D investments are concentrated in therapeutic areas with unmet needs where the risk of failure is very high. With such high attrition rates, to have a good chance of getting at least one successful drug at the end of the process, it is necessary to screen a very large number of molecules.

Furthermore, the expiration of patents on blockbuster drugs, the entry into the market of generic drugs at competitive prices and the increasingly tight budget of the

Fig. 2 Number of New Molecular Entities Approved Per US$ Billion R&D Spend. *Source* Ringel et al. (2020)

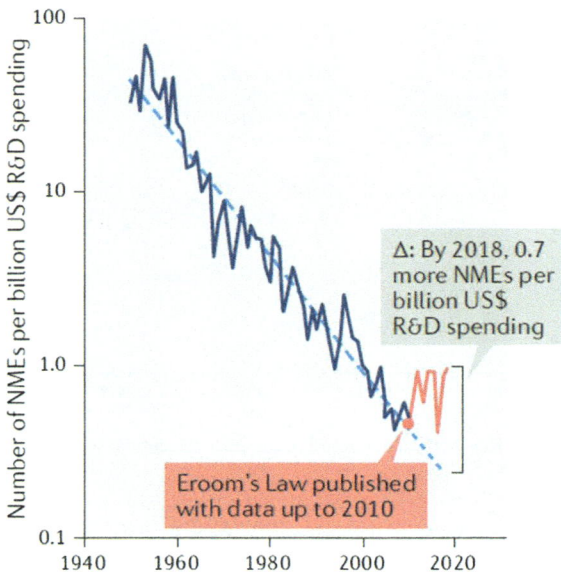

healthcare industry are surrounding conditions that make the challenges for pharmaceutical companies even more difficult and demanding.

Over the past decade, some scholars have noted an inversion of Eroom's law. As of 2010, there have been an additional 0.7 launches of new molecular entities (NMEs) per million dollars of R&D spending per year by 2018. The main reason is the increase in success rates, mainly due to the availability of better information for decisions (Ringel et al. 2020). However, the efficiency of R&D activities is still a key challenge for this industry (Fig. 2).

To outline the challenges the pharmaceutical industry has been facing in recent years, at first the R&D costs trend is analyzed, then the outputs produced, i.e., the number of products in the pipeline, which are considered as a measure of R&D activity.

2.2 Analysis of R&D Costs in the Pharmaceutical Sector

As previously mentioned, the pharmaceutical industry is one of the sectors that boasts the largest investments in R&D. Globally, USD 186 billion was spent on R&D in 2019, for a total of 50 billion more than in 2012 and the trend is strongly growing. Total pharmaceutical R&D spending is estimated to be USD 230 billion

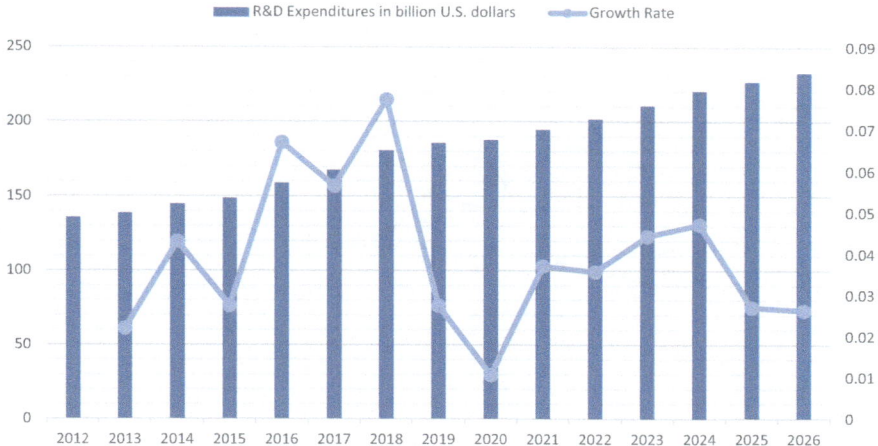

Fig. 3 Total R&D expenditures in the global pharmaceutical industry: historical data and forecasts for the future. Data source EvaluatePharma, 2020

in 2026 (EvaluatePharma, 2020).[1] An increasing use of resources in Research and Development should be an indicator of the greatest commitment of companies in the search for innovations, and therefore be considered positively. However, this is not the case in the pharmaceutical industry, where the greatest investments are the consequence of an increasingly inefficient use of resources (Fig. 3).

Among the major companies are top investors Swiss companies Roche and Novartis, followed by the US Pfizer, Merck and Co., Bristol-Myers Squibb, and Johnson & Johnson (Christel 2021). Roche is the pharmaceutical company that devotes the most resources to R&D (over 14 billion US dollars in 2020) and forecasts for the next few years seem to confirm this hegemony.

To conduct a more in-depth analysis on the trend of R&D costs and on the output of the R&D activity, data relating to 15 pharmaceutical companies were collected, including the global top ten by turnover in 2020 and 5 other smaller companies, but still counted among the Big Pharma, having a total annual production value of more than 10 billion US dollars. They are the most influential companies in the pharmaceutical industry, leading the entire industry and having great power over the global economy. Two clarifications are necessary: Johnson & Johnson is a multinational that does not operate exclusively in the biopharmaceutical sector, but also produces personal care products, self-medications, and medical devices, so the value of production does not refer only to the pharmaceutical sector. Another company could fully fall under Big Pharma, namely the German Boehringer Ingelheim. However, due to the lack of data available, it was excluded from the set of companies.

---

[1] https://fondazionecerm.it/wp-content/uploads/2020/07/EvaluatePharma-World-Preview-2020_0.pdf.

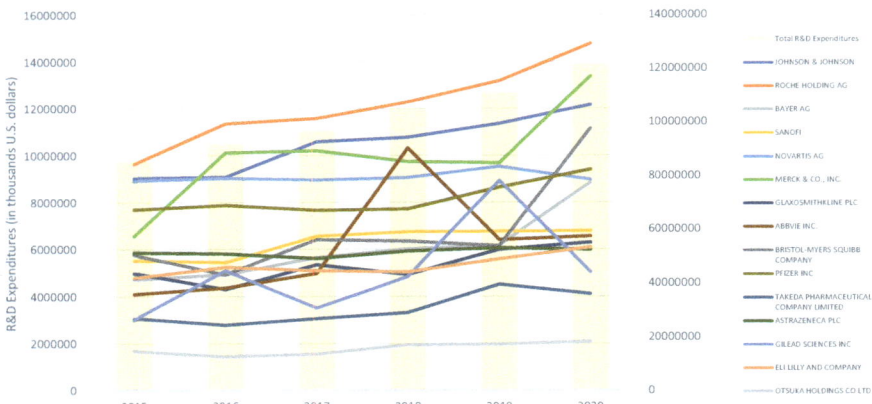

Fig. 4 R&D expenditure of the 15 most important pharmaceutical companies. Data source Bureau van Dijk's Orbis database (some adjustments were needed using data from Yahoo Finance and Macrotrends

As regards R&D costs, data about the 15 companies were collected from 2015 to 2020. As can be seen in Fig. 4, the trend in R&D spending in recent years has largely been increasing, demonstrating that the majority of pharmaceutical companies are raising their investments in R&D in absolute terms year by year.

It is necessary to take into account the complexity of estimating R&D costs, which in some cases could also contain expenses for the acquisition of external R&D projects. For example, Gilead Sciences incurred R&D cost of approximately US$ 5 billions in 2020. However, this figure does not take into account the purchase of IPR&D (In-process R&D), which would raise the total amount of expenditure to around 10 billion.

In addition to observing an overall growing trend in terms of absolute expenditure, it is also important to relate this expenditure to the total value of production. As shown in Fig. 5, in recent years the percentage has remained almost constant for all the companies considered, except for Gilead Sciences and Abbvie which reached a peak in 2019 and 2018, respectively, and then returned to values similar to those of previous years.

The highest percentage is that of Gilead Sciences, which in 2019 invested 40% of the total value of production to R&D expenses.

2.3 Analysis of Pipeline Drugs

A drug pipeline refers to the set of drugs under development by a pharmaceutical company. A pipeline includes products belonging to all the phases, i.e., pre-clinical phase, clinical test phases, regulatory approval phase, and market launch phase. How-

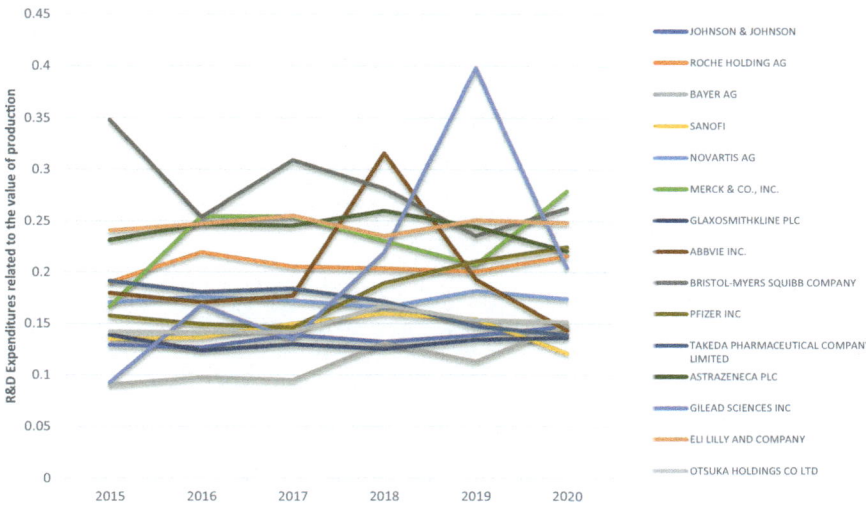

Fig. 5 R&D expenditure as a percentage of the total value of production in the last 6 years. Data source Orbis database, Yahoo Finance, Macrotrends

ever, only those in the post-launch monitoring phase are considered to be part of the pipeline, while drugs whose development has stopped or is complete are typically excluded. The size of a company's drug pipeline is a good indicator of the dynamism of R&D activities. Having a large number of drugs in the pipeline means having more opportunities to obtain successful drugs, thus mitigating the risk of failures. Overall, in recent years, the number of pharmaceuticals in the pipeline of companies around the world has seen and continues to see a slight increase (Informa Pharma Intelligence, 2021),[2] which is, anyway, less than the growth in R&D investments. However, the annual growth rate is highly variable according to the phase considered. Typically, higher growth rates occur in the pre-clinical phase, while the most "stagnant" phases are always phase II and phase III of clinical trials (i.e., those in which the attrition rate is highest). Among the 15 pharmaceutical companies analyzed, there are many differences in terms of pipeline size; among those with the highest average number of drugs in the pipeline is Novartis, which had 232 drugs in the pipeline in January 2021. Lastly, Gilead Sciences, with an average number of 69 drugs in the pipeline in the last 7 years, has recorded a significant increase in the last year (Fig. 6).

---

[2] https://pharmaintelligence.informa.com/~/media/informa-shop-window/pharma/2021/files/infographic/pharmard_whitepaper.pdf.

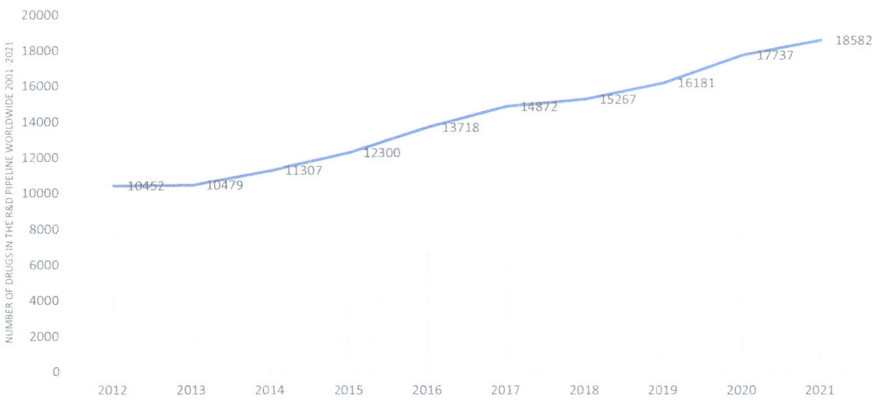

Fig. 6 Evolution of the number of drugs in the pipeline worldwide. Data source Informa Pharma Intelligence, 2021

With regards to the trend in recent years, in Fig. 7 we can observe variable behavior among different companies. The size of the pipeline has always remained close to the average value for Bayer, Otsuka, Eli Lilly, and Novartis. More significant changes were recorded for Takeda, GlaxoSmithKline [GSK], and AstraZeneca. In particular, GSK and AstraZeneca have recorded a sharp decline in the number of drugs in the pipeline in recent years.

In general, there is a positive correlation between R&D expenditures and the average size of the pipeline: companies that spent more resources in R&D over the 6 years maintained an overall higher average number of drugs in the pipeline than those who invested less. However, there are companies that have proved less efficient, since despite spending a lot, they have not managed to keep a sufficiently high number of drugs in development (Fig. 8). The trend of costs and the trend in the number of pipeline drugs can be jointly observed to get an idea of the efficiency of investments in research and development, considering the totality of all 15 companies analyzed. The results are visible in Fig. 9. As can be seen, the total costs are increasing, while the total number of drugs in the pipeline has a downward trend. In 2020, however, a reversal of direction as far as the number of drugs in development takes place.

3 New R&D Open Innovation Models for Pharmaceutical Companies

The imbalance between inputs and outputs has put a question mark over the long-term sustainability of the R&D model of the pharmaceutical sector and has forced large companies to seek solutions that allow them to improve their productivity. On the one hand, there are compelling and established reasons behind the choice of a

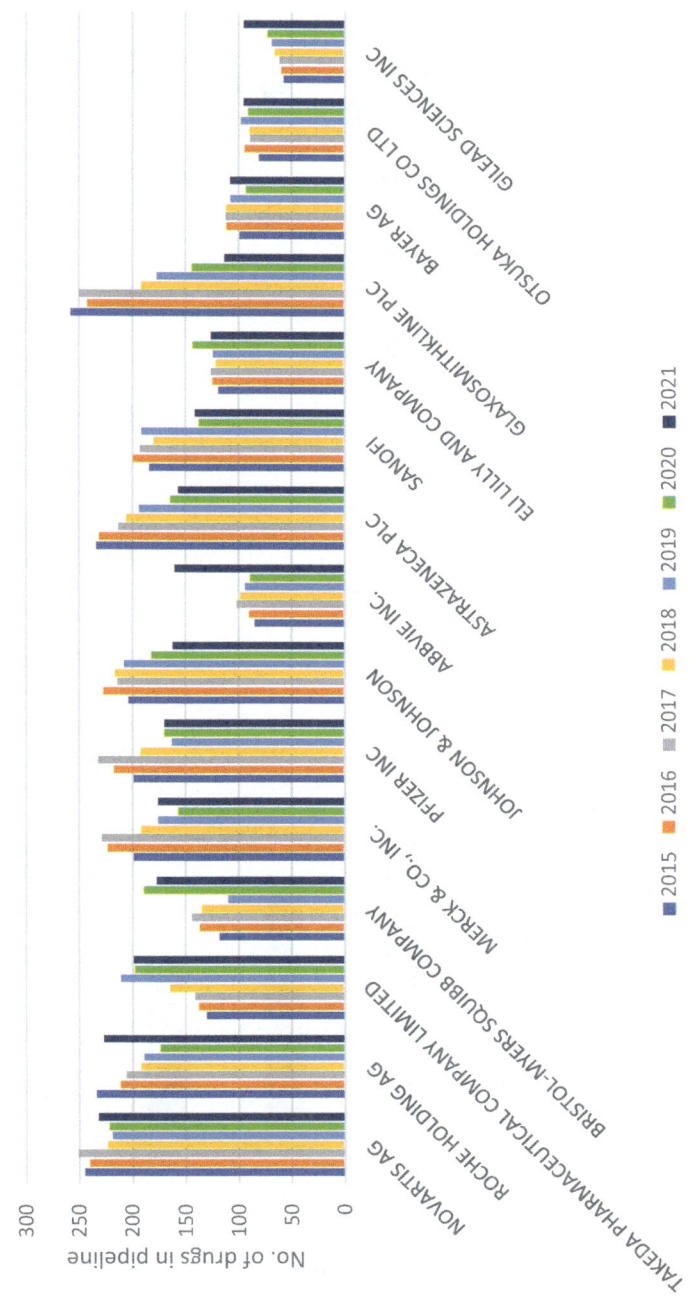

Fig. 7 Number of drugs in the pipeline of the 15 Big Pharma from 2015 to 2021. Data source reports by Pharma Intelligence, available at https://pharmaintelligence.informa.com

Fig. 8 Relation between cumulative R&D expenditure (2015–2020) and average pipeline size (2015–2021). *Source* Pharma Intelligence (https://pharmaintelligence.informa.com) and Bureau van Dijk Orbis

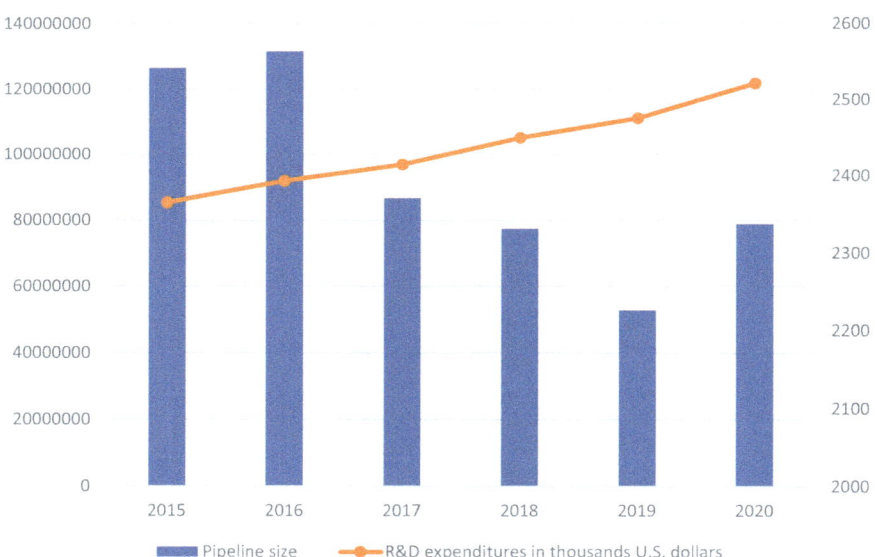

Fig. 9 Trend of the number of drugs in the pipeline (secondary axis) and R&D expenses in thousands USD (primary axis) of the 15 pharmaceutical companies. *Source* Pharma Intelligence (https://pharmaintelligence.informa.com) and Bureau van Dijk Orbis

closed R&D process—which basically consists in preventing the disclosure of all the important information to the outside. The most notable motivations are the fear of the competition (that could use this information to be the first in the market), the need to protect the intellectual property, and the confidentiality of the patients' data (Au 2014). On the other hand, this closed model did not prove to be the most effective because of the evidence previously provided. This is critical for all the companies in the industry, which are impelled by the imperative of continuously innovating and staying competitive. Therefore, all the pharmaceutical companies are changing their business models, in particular their strategies for R&D, moving toward more "open" models, where the knowledge is shared between different actors. The newly developed business model features collaborations with a variety of actors, including academic research institutions, private and public organizations, and other companies operating in the healthcare environment at every stage of the R&D process.

In recent times and with ever-increasing intensity, Big Pharma is among the main players in transactions such as mergers and acquisitions [M&A], partnerships and collaborations, and outsourcing of products and processes, which involve a complete restructuring of the R&D. Following the example of other industries which pioneered the sharing of knowledge, pharmaceutical multinationals have started to realize the full potential of open innovation. Companies can therefore acquire products, processes, or entire companies to fill any gaps in their product portfolio or simply to enrich their pipeline of candidate drugs, or they can access external know-how in outsourcing, to broaden their knowledge and contribute to the discovery of new compounds. Also, FDA and EMA decided to give free access to their clinical trial database in order to let anyone use the data to help accelerate the new drug development process and make the approval process more efficient (Au 2014).

The open innovation in the pharmaceutical industry is linked to the important concept of "absorptive capacity", which is defined as a firm's "ability to recognize the value of external knowledge, assimilate it, and apply it to commercial ends" (Cohen and Levinthal 1990). The new business models promoted by pharmaceutical companies are based on this ability. Pharmaceutical companies benefit from absorbing external knowledge from other practitioners in the field in order to best adapt to the latest technologies and innovations (Romasanta et al. 2022). Patterson and Ambrosini (in Patterson and Ambrosini (2015)) defined the absorption process of the pharmaceutical industry as follows: search and recognize the value of new knowledge (through connections with universities, research organizations, and so on), assimilate it (namely, process and interpret it), acquire the rights of using this knowledge, transform the knowledge further developing it and combining it with the existing knowledge of the company and at the end exploit this knowledge to obtain important results (Patterson and Ambrosini 2015). In general, pharmaceutical companies have several options available to access external knowledge. They can apply traditional strategies such as M&A, licensing agreements, and partnerships with research institutes or universities, in order to strengthen the internal R&D efforts, and/or take advantage of opportunities throughout the R&D value chain to access external sources of innovation. Open innovation models are increasingly popular: they consist of new ways of carrying out R&D activities within the organizations, actively involving external actors, whether they are experienced researchers from academic institutions or scientists. A growing number of pharmaceutical companies are turning to these alternative models of open innovation, breaking down the barriers between the internal organization and the environment that surrounds it, populated by a large number of actors that can bring significant value to the company. The companies are required to integrate different strategies, both traditional and innovative, so as to build new and more efficient R&D models. Currently, the standard for pharmaceutical multinationals consists in having a portfolio mainly containing externally generated projects (Schuhmacher et al. 2013). Two different types of R&D partnerships, related to different kinds of resource outsourcing, can be identified:

- Outsourcing of products within different stages of the R&D process or of entire processes, through license agreements or M&A (direct partnerships).
- Outsourcing of knowledge, consisting in the integration of external knowledge within the organization through collaborations, partnerships, and open knowledge platforms (indirect partnerships).

Direct outsourcing, obtained through direct partnerships, brings more immediate benefits, such as a rapid increase of drug candidates in the pipeline, but it is more expensive and therefore riskier. Furthermore, through direct outsourcing the company could risk losing its ability to innovate, excessively relying on external sources. Indirect outsourcing, via indirect partnerships, can lead to greater difficulties in the immediate term, as integrating internal and external knowledge is not always easy, but in the long term it can bear fruit at much lower costs than direct outsourcing (Fig. 10).

The models included in both the categories of direct and indirect partnerships will be described.

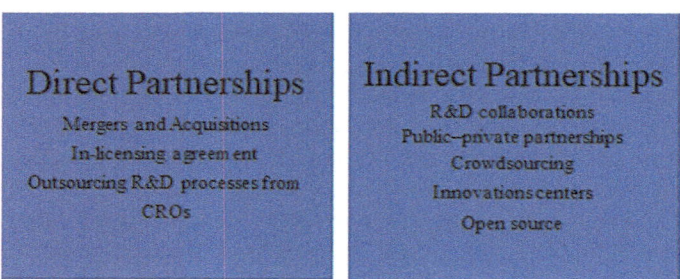

Fig. 10 Direct and indirect partnerships.eps

3.1 Mergers and Acquisitions

The pharmaceutical sector has always been characterized by an intense M&A activity, in which large multinationals acquire smaller companies with the aim of increasing size and quality, as well as diversifying their drug pipeline. Through these types of operations, pharmaceutical companies seek to promote better financial and operational performance. The term M&A refers to transactions that take place between two entities that combine in some way. Even if, usually, the two terms "merger" and "acquisition" are used interchangeably, they refer to two distinct operations. In a merger, two companies of similar size merge to form a single entity, while an acquisition takes place when a larger company buys a smaller company, completely absorbing it. Both types of operations can be hostile or friendly.

The most common reasons why pharmaceutical companies frequently resort to this type of transaction are the following:

- Offset the loss of revenue due to the expiration of blockbuster drugs and the entry into the market of generic drugs;
- Expand its research scope to include new therapeutic areas;
- Access patented technologies of strategic importance;
- Expand the drug pipeline;
- Enter new markets.

In the pharmaceutical industry, M&A transactions are usually of the horizontal type. It means they take place between companies both producing drugs, but which differ, for example, by therapeutic areas treated or by type of drugs developed. The most popular acquisitions are those that large pharmaceutical companies operate toward small biotech companies, in order to include in their R&D project portfolio, not only drugs obtained from chemical or artificial products, but also drugs produced from living organisms. Pharma Intelligence (2020) reports that from 2010 to 2019 there were over 700 M&A transactions in the pharmaceutical and biotech industries, with Abbvie, Takeda, and Bristol-Myers Squibb in the top three by number of agreements signed. Indeed, in 2018, Takeda acquired Shire for $62 billion, while in 2019, Bristol-Myers Squibb acquired Celgene ($74 billion) and Abbvie Allergan (for $63 billion) (Statista, 2021).[3]

3.2 In-licensing Agreements

Pharmaceutical companies can acquire the right to dispose of a drug patented by another company (or a research laboratory) through a licensing agreement, in exchange for paying royalties to the licensor company. Pharmaceutical and biotechnology companies are turning to the in-licensing agreements with increasing frequency. This leads to considerable advantages on both sides: the in-licensing allows

---

[3] https://www.statista.com/statistics/518674/largest-mergers-acquisitions-pharmaceutical/.

pharmaceutical companies to enrich their pipeline and to market new drugs in authorized countries. On the other hand, the advantage for the biotechnology companies is that they can access the resources needed for the development of the final stages of the development process, for clinical trials, for the production and distribution, sharing the risk with companies that are larger and more solid from a financial point of view. Today many of the candidate drugs included in the big pharma pipeline are not developed in-house, but outsourced to other companies, typically smaller and more research-focused, just like biotechnology companies. Pharmaceutical companies that own most internally developed drugs are commonly referred to as "research-based". A drug obtained via a licensing agreement can be in different stages of development, so it can also be ready to market. In this case, the license covers only the sale of the drug. If the drug is still under development, the company that purchased the license will have to take care of the continuation of this process. The achievement of certain milestones in development can involve the payment of additional royalties to the licensor.

In some respects, licensing agreements are much more convenient compared with the traditional drug discovery process, in which a company embarks on a project by investing large sums and taking a very high risk. Furthermore, in-licensing is more attractive than M&A operations. This is because the licenses allow pharmaceutical companies to purchase the rights to experimental drugs without having to carry the entire baggage of another organization. According to an analysis carried out by KPMG (2021),[4] 593 licensing agreements were signed between pharmaceutical companies around the world in 2020, with a substantial growth compared to the 360 of the previous year. Among the top deals of 2020 we find the $1.7 billion worth deal between Roche and Blueprint. With this agreement, Roche has added Blueprint medicine RET inhibitor pralsetinib to its cancer drug portfolio. Another remarkable licensing agreement is the one signed by AstraZeneca and Daiichi Sankyo for one of the Japanese pharmaceutical company's antibody-drug conjugates (ADCs). Further agreements were signed by Sanofi, Merck and Co., Abbvie and Eli Lilly (Taylor, 2021).[5]

3.3 Outsourcing R&D Processes from CROs

Contract Research Organizations [CROs] are organizations that operate in the service sector providing support to pharmaceutical and biotechnology companies. The latter can outsource research services for new drugs or medical devices, avoiding the huge costs that these processes would require if carried out internally. Research services range from drug discovery to commercialization, with a particular focus on the clinical phase. Rather than hiring permanent employees with the specialist knowledge

---

[4] https://advisory.kpmg.us/content/dam/advisory/en/pdfs/2021/biopharmaceuticals-deal-trends.pdf.

[5] https://www.fiercebiotech.com/special-report/top-15-biopharma-licensing-deals-2020.

necessary to perform some of the activities included in the development process in-house, pharmaceutical companies can pay a CRO to carry out these activities externally. The CRO, acting as an independent contractor with specialist knowledge, performs a number of tasks to help complete the new drug development process faster, more efficiently, and at a lower cost. CROs can be large multinationals or small specialized companies. Leading Contract Research Organizations worldwide include the Laboratory Corporation of America Holdings, IQVIA, Syneos Health, and Pharmaceutical Product Development. The global CRO market in 2015 began a period of unstoppable growth, reaching a total value of over 40 billion US dollars in 2020 which is estimated to exceed 60 billion by 2024 (HKExnews, 2020).[6]

3.4 R&D Collaborations

R&D collaborations belong to the category of indirect partnerships, as they do not involve the direct purchase of outsourced products or services from a third party, but instead the sharing of knowledge that leads to the development and marketing of new drugs. In the pharmaceutical industry collaborations with third parties are aimed to access specialized know-how, whether it is from other companies, such as biotechnology companies, or from entities belonging to the academic world, such as universities and private or public institutions. Given the growing complexity of the R&D of new drugs, today, collaborations have increased considerably, creating real ecosystems in which knowledge can be shared much more simply. Collaborations are used to access knowledge, concerning for example new drug targets, biomarkers, animal models, and translational medicine. Alliances, collaborations, and partnerships can bring substantial improvements in clinical success, while reducing long development times and total expense. Through them, each company can maintain only its own core competencies and rely on third parties just to acquire knowledge regarding areas of non-competence. Indeed, in recent years, R&D leaders have realized that by maintaining the total depth and breadth of research projects, the totality of activities cannot be done through internal efforts alone. Numerous pharmaceutical companies have moved R&D locations close to the finest academic institutions. One example is Pfizer, which in 2014 opened a new research facility in Cambridge (Massachusetts), where two of the world's most famous academic institutes are based, Harvard University and the Massachusetts Institute of Technology (MIT). The advantages are reciprocal: pharmaceutical companies can benefit from the flows of knowledge promoted by the academic environment and academic partners can count on the economic support of the large pharmaceutical multinationals. In addition to Pfizer, GSK and Merck & Co. have also initiated massive collaborative operations (Schumacher et al. 2016). Roche Holdings official website writes: "As a pioneer in healthcare we are committed to driving groundbreaking scientific and technological advances that have the potential to transform the lives of patients worldwide. But we can't accomplish this on our

---

[6] https://www.statista.com/statistics/813502/market-size-clinical-cro-worldwide/.

own. Only by partnering with the brightest minds in science and healthcare can we serve the needs of patients" (Roche Website). On the same page, Roche reports some useful data to understand the company's effort in enforcing external collaborations: 40% of total sales come from a partner or in-licensed products, 50% of the pipeline comes from outside the organization and there are 220 active global partnerships. Over time, this has certainly contributed to making the company one of the world leaders in the industry.

3.5 Public-Private Partnerships

Public-Private Partnerships (PPPs) consist of collaborations between public organizations and private companies with the purpose of carrying out different projects. Typically, these agreements are used to finance projects such as the construction of public transport networks, and parks. In the pharmaceutical industry, these agreements provide for the financing of R&D activities through public funds or charities aimed at discovering new drugs, especially in therapeutic areas where there are unmet medical needs and where pharmaceutical research is less active. An example of a successful PPP is IMI [Innovative Medicines Initiative], a public-private partnership between the European Union and the European Federation of Pharmaceutical Industries and Associations [EFPIA], which aims to speed up drug development, making it better and safer, especially in areas of particular need. IMI creates a real network of experts to promote pharmaceutical innovation in Europe, bringing together universities, small- and medium-sized enterprises, pharmaceutical multinationals, research centers, patient organizations, and regulatory authorities (European Commission, 2020).[7] Funding comes from both the European Union and the EFPIA. From 2004 to 2020, funds were raised for a total of over 3 billion euros (IMI Europe). EFPIA member companies do not receive money through IMI but contribute to the partnership by providing staff, funds, clinical data, samples, compounds, etc. The beneficiaries of these resources are universities, research organizations, small- to medium-sized companies, and all the entities listed above, which work to achieve common goals. Among the most famous multinationals, Sanofi, GSK, J&J, and AstraZeneca are the ones that contributed the most, providing several teams working on numerous projects. Public-private partnerships are an open innovation model that helps companies improve business competitiveness and reduce R&D costs. Furthermore, they mitigate competition and help reduce the fragmentation of knowledge typical of the pharmaceutical sector (Schumacher et al. 2018).

---

[7] https://ec.europa.eu/info/research-and-innovation/research-area/health-research-and-innovation/innovative-medicines-initiative_en.

3.6 Crowdsourcing

The term crowdsourcing refers to a type of activity carried out online in which a company, an organization, an institution, or a private individual proposes a problem and requests opinions, suggestions, and ideas to solve it from internet users. It is a valid economic model for the development of projects, which promotes the meeting between supply and demand. Crowdsourcing has become a very common practice in the pharmaceutical industry: companies create platforms on which to post problems and questions, inviting all experts to find solutions in exchange for a cash prize. The basic concept is always that of the absorption of knowledge from the external environment, as in partnerships and collaborations. Crowdsourcing is an extremely open, highly innovative, and inexpensive model that brings different benefits to the organizations. Through it, pharmaceutical companies can have access to a broad solution diversity and specialized skills can mitigate the risk that competitors would exploit the same openness and increase the brand visibility (Christensen and Karlsson 2019). Sanofi, Eli Lilly, Bayer, and AstraZeneca are just some of the companies that have fully exploited the potential of crowdsourcing by creating specific platforms, directly accessible from their websites. On these platforms, it is also possible to access a large set of data to deepen one's knowledge on human diseases, collections of targeted molecules, and compounds at various stages of the development process. In this way, a bilateral sharing of knowledge is created and innovation is easily promoted. One of the best-known Open Innovation platforms is InnoCentive,[8] a global network where experts from different industries propose solutions to problems posted by public and private companies, government institutes, non-profit organizations, research institutes, and public and private laboratories. InnoCentive is a startup born from the multinational Eli Lilly, with the aim of seeking outside the company solutions to problems that internal experts were unable to solve. Among the most well-known challenges is that of TB Alliance, a non-profit organization that deals with discovering and developing new economic drugs for the treatment of tuberculosis. To involve as many people as possible, the TB Alliance proposed a contest on InnoCentive in which it sought a solution to reduce the process cost of an existing drug. Ultimately, the solution was found by two scientists, benefiting millions of people with the disease, particularly in developing countries. Other companies have, instead, a more cautious approach to crowdsourcing. They promoted initiatives like challenges, calls for tenders, and real contests. It is the case of Roche, Takeda Italia, and Pfizer.

---

[8] www.innocentive.com.

3.7 Innovation Centers

Another model of open innovation is innovation centers, i.e., real centers that bring together scientists belonging to pharmaceutical multinationals and academic experts from all over the world, with the aim of promoting intense collaborations that can lead to innovative solutions. This model can be considered as a hybrid between a centralized R&D model with elements of open innovation (Schumacher et al. 2018). Merck has built its innovation center in Darmstadt, Germany, creating a welcoming and inspiring environment where both employees and partners can grow their ideas. An academy is also held in the center, where talented young people are guided in a training process aimed at building solid and profitable knowledge. Roche owns 7 innovation centers around the world (Zurich, Shanghai, New York, Munich, Basel, Copenhagen, Welwyn) where the sharing of skills, information, and technologies is encouraged. J&J also boasts 4 innovation centers in San Francisco, Boston, London, and Shanghai, where integrated teams of experts continuously collaborate with entrepreneurs and scientists from around the world contribute to progress in the pharmaceutical field. Innovation centers, therefore, form ecosystems in which knowledge can circulate more quickly and efficiently, bringing numerous benefits to all the players involved and promoting scientific progress on a global level.

3.8 Open Source

Open source is a widely used model in the software industry, where it has gained tremendous success in the last few years. The model consists in making source code, blueprints, or documentation available free of charge to anyone, so that it can be further developed or modified and then shared with the community. Open source is therefore based on transparency, it promotes free access to knowledge to obtain collaborative improvements without any financial compensation in return. All these concepts seem to clash with the general principle of the pharmaceutical industry, which is rather closed and highly competitive. However, the ultimate goal of pharmaceutical companies should be to create drugs that can improve people's living conditions, and not to make the highest profits possible. Therefore, open-source models would fit well with the principles that, in theory, the pharmaceutical industry should adhere to. The issue is actually more complex, because if the information on the development of new drugs was shared without any form of protection, pharmaceutical companies would hardly be able to obtain profits so as to recover research costs and, therefore, would be discouraged and probably R&D would suffer a major setback. For this reason, open source in the pharmaceutical sector has not had the same success as in the software industry. However, several open-source initiatives have been promoted by pharmaceutical companies such as GlaxoSmithKline, which together with MIT and Alnylam Pharmaceuticals, formed the "Pool for Open Innovation against neglected tropical diseases", which allows free access to 2,300 patents on drugs that treat tropical diseases. Another example is the "Drugs for Neglected Diseases Initiative", a free platform on which information on the discovery and development of drugs for the treatment of diseases such as sleeping sickness, pediatric HIV, and Chagas disease is shared by AstraZeneca, Bayer, Bristol-Meyers-Squibb, Novartis, GSK, Pfizer, Sanofi, and Takeda (Schumacher et al. 2018). Open source in the pharmaceutical field is therefore mainly used to promote the development of drugs intended for the treatment of widespread diseases, especially in developing countries.

References

Au, R. (2014). The paradigm shift to an "open" model in drug development. *Applied & Translational Genomics, 3*(4), 86–89 (Global Sharing of Genomic Knowledge in a Free Market). https://doi.org/10.1016/j.atg.2014.09.001, https://www.sciencedirect.com/science/article/pii/S221206611400026X

Christel, M. (2021). Pharm exec's top 50 companies. *Pharmaceutical Executive* 406. https://www.pharmexec.com/view/pharm-execs-top-50-companies-2020

Christensen, I., & Karlsson, C. (2019). Open innovation and the effects of crowdsourcing in a pharma ecosystem. *Journal of Innovation & Knowledge, 44*, 240–247. https://doi.org/10.1016/j.jik.2018.03.008, https://www.sciencedirect.com/science/article/pii/S2444569X18300362

Cohen, W. M., & Levinthal, D. A. (1990). Absorptive capacity: A new perspective on learning and innovation. *Administrative Science Quarterly, 35*(1), 128–152. http://www.jstor.org/stable/2393553

Patterson, W., & Ambrosini, V. (2015). Configuring absorptive capacity as a key process for research intensive firms. *Technovation, 36–37*, 77–89. https://doi.org/10.1016/j.technovation.2014.10.003, https://www.sciencedirect.com/science/article/pii/S0166497214001382

Ringel, M. S., Scannell, J. W., Baedeker, M., & Schulze, U. (2020). Breaking Eroom's law. *Nature Reviews Drug Discovery, 19*(12), 833–834. https://doi.org/10.1038/d41573-020-00059-3, https://pubmed.ncbi.nlm.nih.gov/32300238/

Romasanta, A. K., van der Sijde, P., & de Esch, I. J. (2022). Absorbing knowledge from an emerging field: The role of interfacing by proponents in big pharma. *Technovation, 110*, 102363. https://doi.org/10.1016/j.technovation.2021.102363, https://www.sciencedirect.com/science/article/pii/S0166497221001449

Scannell, J.W., Blanckley, A., Boldon, H., & Warrington, B. (2012). Diagnosing the decline in pharmaceutical r&d efficiency. *Nature reviews Drug Discovery, 11*(3), 191–200. https://doi.org/10.1038/nrd3681, https://pubmed.ncbi.nlm.nih.gov/22378269/

Schuhmacher, A., Germann, P.G., Trill, H., & Gassmann, O. (2013). Models for open innovation in the pharmaceutical industry. *Drug Discovery Today, 18*(23), 1133–1137. https://doi.org/10.1016/j.drudis.2013.07.013, https://www.sciencedirect.com/science/article/pii/S135964461300247X

Schuhmacher, A., Wilisch, L., Kuss, M., Kandelbauer, A., Hinder, M., & Gassmann, O. (2021). R&d efficiency of leading pharmaceutical companies—A 20-year analysis. *Drug Discovery Today, 26*(8), 1784–1789. https://doi.org/10.1016/j.drudis.2021.05.005, https://www.sciencedirect.com/science/article/pii/S1359644621002361

Schumacher, A., Gassmann, O., & Hinder, M. (2016). Changing R&D models in research-based pharmaceutical companies. *Journal of Translational Medicine, 14*(1), 479–5876. https://doi.org/10.1186/s12967-016-0838-4

Schumacher, A., Gassmann, O., & Hinder, M. (2018). Open innovation and external sources of innovation. an opportunity to fuel the R&D pipeline and enhance decision making? *Journal of Translational Medicine, 16*(1), 479–5876. https://doi.org/10.1186/s12967-018-1499-2

Gianpaolo Iazzolino Associate Professor in Business Economics at the Department of Mechanical, Energy and Management Engineering at the University of Calabria, Italy. His main research interests are in Firm Performances and Evaluation of Intangibles. He is currently Delegate of the Rector for the Right to Education and member of the Technical Committee for spin-offs at the University of Calabria.

Rita Bozzo Received her M.Sc. in Management Engineering (cum laude) from the University of Calabria in 2021 with a thesis on the R&D processes in the pharmaceutical sector. She is currently employed at a company operating in the automotive industry with a focus on connected mobility.

Open Access This chapter is licensed under the terms of the Creative Commons Attribution 4.0 International License (http://creativecommons.org/licenses/by/4.0/), which permits use, sharing, adaptation, distribution and reproduction in any medium or format, as long as you give appropriate credit to the original author(s) and the source, provide a link to the Creative Commons license and indicate if changes were made.

The images or other third party material in this chapter are included in the chapter's Creative Commons license, unless indicated otherwise in a credit line to the material. If material is not included in the chapter's Creative Commons license and your intended use is not permitted by statutory regulation or exceeds the permitted use, you will need to obtain permission directly from the copyright holder.

Modelling Specific Business Processes in the Life Science Industry

Pharma Tender Processes: Modeling Auction Outcomes

Philipp Mekler and Jingshu Sun

1 Introduction

Tendering is a commonly used process where government and public institutions grant bidding opportunities for large projects with a defined bidding target and a defined bidding date. Pharmaceutical tenders represent the process when a number of manufacturers offering the price of similar or comparable products to win the privilege to gain the monopsony in the market. During the confidential bidding process, the winning party is offered the opportunity to sell the bidding product at a pre-defined price for a fixed period of time, which reflects the nature of tendering as "winner takes it all", as stated by Petrou (2016). From the contract or mechanism design theory perspective, tendering process can be linked to auction theory in view of competition. Thus auction theory is widely used in modeling the price competition during the tendering process. From the social welfare perspective, according to Simoens and Cheung (2019), the tendering process encourages competition and thus cuts the price, which is significantly beneficial to yield short-term savings, especially under extreme budget constraints. In the literature of pharmaceutical tendering, there are two main branches of research: The first stream provides qualitative descriptions on tendering and procurement systems of different countries and their impact, both short-term and long-term, on drug price and market concentration. In addition to overall nationwide impact, some researchers also examine the determinants of bidding behaviors. The second stream focuses on the empirical estimation of the best winning bidding prices in pharmaceutical tenders.

P. Mekler
University of Basel, Basel, Switzerland
e-mail: philipp.mekler@unibas.ch

J. Sun (✉)
F. Hoffmann-La Roche AG, Basel, Switzerland
e-mail: jingshu.sun@roche.com

This review focuses on two parts: an overview of tendering systems and elements, and a summary of empirical methods to predict the best bidding price. It will start with an introduction of pharmaceutical bidding in Sect. 1. Then in Sect. 2, the tendering system in four different countries is discussed qualitatively and the impact on local pricing and market landscape is evaluated. Section 3 follows with a definition of the scope of pharmaceutical bidding from auction theory. This includes the common auction elements such as auction time, information asymmetry, and players. Next, in Sect. 4 we summarize the two most commonly used empirical methods for price auction prediction. In this section, we also cover recently involved testing methods for price estimation. Section 5 provides thoughts on different empirical models and proposes best practice to select models based on experimental data.

2 Pharmaceutical Tendering

Pharmaceutical tendering is a common procurement process across countries but could be regulated quite differently depending on the healthcare system of each country. Maniadakis et al. (2018) points out that several factors contribute to these differences, including demographic dynamics, economic growth, and distribution of public bodies. In this section, we summarize the pharmaceutical tendering market from four different countries, both at nation level and at the state or province level. We select the following four countries from four different continents to reflect the diverse nature of pharmaceutical tendering process with different regulatory set up and economic situations. We also provide an overview of the impact of pharmaceutical tendering on drug price and market dynamics.

2.1 Pharmaceutical Tendering Mechanism in Different Countries

In this subsection, we summarize the public pharmaceutical tendering system in different countries. Most countries follow the Public Procurement Act at the country level, while for other large countries such as Brazil and China, procurement for pharmaceutical auctions is organized at state or province level.

2.1.1 Tendering Mechanism in Sao Paulo, Brazil

In Brazil, the Public Procurement Act effective 1993 regulates the procurement process of public goods, including all pharmaceutical procurements, as illustrated by Paulo et al. (2013). The Act requires that all public bodies show a clear description with detailed documentation of the input good, bidding quantity, product quality, origin, and delivery details. According to Paulo et al. (2013), the Brazil legislation specifies two forms for standard purchasing inputs: physical submission and electronic reverse auctions. Paulo et al. (2013) also point out that within the physical submission, there are two types of auctions, open competitive bidding and invited bids. Within electronic auctions, there are three types, first-price sealed bid auction, English auction, and two-stage auction. There are some additional auction requirements, for example, high-value contracts can only be completed through open competitive bidding. Sao Paulo has to follow additional rules due to its large population, unique economic contribution, and complex industrial situation. First, all the procurements in Sao Paulo have to be placed via the common electronic platform, namely, only in the form of an electronic reverse auction. In terms of auction type, there is also a difference where only first-price sealed bid auction or two-stage auction is allowed. The auction type is determined by the value of contracts. Low-value contracts are acquired through first-price sealed bid auctions. The first-price auction is achieved in two steps in Sao Paulo: in the pre-bidding stage, the auction notice is available for five days with detailed description of bidding information, terms, and conditions; in the bidding stage, the bidder submits one single sealed bid before the deadline without knowing the identity of the other bidders. According to Paulo et al. (2013), in Sao Paulo, public hospitals, health agencies, and medical centers obtain prescription drugs in a de-centralized way, in which each entity is responsible to acquire what is needed. These public bodies have to rely on the auction-based tendering process to acquire those generic drugs in Sao Paulo.

2.1.2 Tendering Mechanism in South Africa

The Public Procurement Act in South Africa has been effective since 1982. The scope of procurement by the national government includes the essential drugs in the public healthcare system. According to Wouters et al. (2019), those essential drugs are further divided into 15 different categories and for each tender, the effective period ranges from two to three years. Market authorization is needed to access the South African tendering system. Both local manufacturers and international suppliers are allowed to participate in the bid but in most situations, international suppliers participate through their locally registered subsidiaries and offices. Different from the first-price sealed bid auction mechanism adopted by many countries, South Africa uses a two-stage scoring system to determine the winning supplier. In the first stage, the manufacturer who offers the lowest price among bidders gets 90 points while at the same time remaining suppliers get deductions in proportion to the difference to the winning offer, which is calculated by a publicly announced method. Then, in the second stage, the 10 points left are divided based on an empowerment score, also called "broad-based black economic empowerment score". This score is decided by the government with a set of criteria, including the diversity of equity owners and managerial roles. When granting the winning bid, price is not the only determinant. Wouters et al. (2019) mention that other ad hoc factors are also considered. For example, under consideration of local economic growth protection, the government may also accept a price mark up to 10 points for national suppliers. In addition to economic factors, industry diversity, labor market balance, and trade situation are also considered. Thus, to keep a proper balance between demand and supply, the government even splits the winning bidding offer between multiple firms if the products are not clearly differentiated. Gray and Suleman (2014) summarize that the bidding process in South Africa is highly regulated and less price originated compared to other countries.

2.1.3 Tendering Mechanism in Cyprus

Petrou (2016) points out that the pharmaceutical market in Cyprus is highly fragmented between the public and private sector, where the public sector has strong regulations on supply and demand. In contrast, regulations only apply to the private market in terms of the price. The wholesale price in the private section can be significantly higher than the official price, where pharmacies add a large mark up. Tendering helps lower the price on both generic and branded products, up to 95 percent of the price for a generic drug and up to 80 percent of the price for a branded drug. Petrou and Talias (2014) state that, however, until 2017, Cyprus was the only country in the EU that did not have a universal coverage national health system (NHS). Petrou and Talias (2015) further state that all the drugs for public healthcare are procured by the Ministry of Healthcare (MoH) by tendering.

2.1.4 Tendering Mechanism in Guangdong, China

Yao and Tanaka (2015) state that Pharmaceutical tendering in China is also implemented at a province level due to the different economic situations across the nation. In Guangdong province, manufacturers can bid for multiple heterogeneous products and complete multiple bids simultaneously. This allows suppliers to combine multiple products together during the bidding process, contrary to the traditional single-product bidding. However, this mechanism is different from package bidding, where bidders can take advantage of the preferred items for higher revenue. In Guangdong, multiple bidding items are allowed but suppliers have to submit the bids separately for each kind of drug. Although the bidding is operated at province level, certain procurement rules still have to be followed at a national level, for example, the rules for drug listings. In China, the central government and provincial municipalities issue multiple layers of drug listings that can be procured, further divided to essential drugs, list A drugs and list B drugs, as introduced by Yao and Tanaka (2015). The above-mentioned categories differ in terms of reimbursement rules and pricing policies. Thus, the price of drugs in China is determined by combining market-driven elements with government regulations. In terms of bidding rules, in Guangdong province, a four-step bidding process is implemented through an online platform. Yao and Tanaka (2015) further explain that: In the first steps, the government screens the access authorization of all interested firms. The Good Manufacturing Practice

(GMP) and Good Supply Practice (GSP) certificates are the prerequisites for bidding participation. In the second step, the bidding drugs are classified into one-shot and three-round bidding drugs. Only essential drugs can be procured in the tendering process. One-shot bidding is used for emergency products, low-priced drugs, and drugs for rare diseases or controlled by governments. Compared to three-round drugs, the competition on one-shot drugs is lower. In the third step, all the participating suppliers offer the bidding price in the pre-procurement stage. If the bidding product is a controlled product, then the winner is already secured in this stage. Otherwise, the bidding needs to go through the three rounds of bidding. In the last step, the participants are bidding on the final outcome. Unlike the first-price sealed bid auction where the lowest price bidder wins all, in Guangdong, the process is operated in three rounds in which the highest price bidder is removed from the game after each round. In each round, the supplier is able to set and change the price according to the bidding results from the previous rounds. In addition, in each round a quota is set and short-listed suppliers will stay in the game until the final quota is reached. Thus, the bidding process in Guangdong province China is much more complicated than a normal first-price auction bid.

2.2 Effect of Pharmaceutical Tendering

There has been a lot of discussion on the effectiveness of pharmaceutical tendering system and its impact on individuals and society. Two substantive areas that have been empirically examined are price competition and market concentration.

2.2.1 Pharmaceutical Tendering and Price Competition

Many studies have empirically examined under the pharmaceutical procurement process, and how the price of drugs and volume of bid change in response to increased competition. In the literature, the effect on price is evaluated separately by product attribute, which is divided into generic drugs and brand-name drugs (also called branded drugs). Brand-name products refer to the drug that is originally developed by a pharmaceutical company, approved by the authority for market access, and sold exclusively under brand name for a fixed time period under patent protection. After the expiration of patency, the branded drug becomes a generic drug that contains the same amount of active ingredients as other generic drugs. The key difference between a branded drug and a generic drug is the exclusive patent protection mechanism and this exclusive right significantly changes the manufacturer's ability to manipulate the price. This is the reason the effect of pharmaceutical tendering is examined separately for these two kinds of drugs.

Petrou (2016) assesses the long-term effect of tendering on price for branded products in the Cyprus market and finds superior price reduction effect through tendering. Using continuous 7-year data from 2006 to 2012 with 36 branded products,

Petrou (2016) adopts the repeated measures generalized linear model and empirically proves that tendering retains its capability in reducing the initial drug price [?]. In addition, the downward trend on price exhibits a continuous trajectory. Technically, within the repeated linear model the paper also includes other potentially explanatory variables such as interchangeability, indication, administration status, sales volume as well as triple interaction terms to account for any variances. Finally, Petrou (2016) introduces Greenhouse-Geisser correction to adjust for time-variant elements in price reduction and concludes that tendering shows a significant long-term price reduction effect and the effect is further moderated by product attributes including interchangeability, in-patient status, and indication.

Wouters et al. (2019) also investigate the long-term effect of tendering for medicines and find similar results based on a 14-year period of tendering data for South Africa but focusing on generic drugs. Computing three different types of price index (Laspeyres, Paasche, and Fisher indexes) across multiple medicine categories, the paper finds that the price in most medicine categories drops consistently over time and the medicine procured through public systems is always lower than the ones from private systems (Wouters et al. 2019). Tendering in general is quite effective in securing a lower drug cost.

While most research focuses on the tendering effect either on generic drugs or branded drugs, Paulo et al. (2013) study the interaction of drug entry on drug price between the two types. The authors identify the causal effect of a generic drug's entry on the bidding participation rate of branded drug manufacturers and the consequences on price paid for the bid. They examine the following three questions: how branded suppliers' participation decision will change in reaction to the presence of a generic drug entry; how the paid bid price would change after a generic supplier appears in the competitive bidding setting; and whether there is a statistical difference on bids and price bid between generic and branded pharmaceutical manufacturers. By using the Brazil transactional procurement data with 30448 records from 2008 to 2012, Paulo et al. (2013) analyze 3859 different drugs from 425 active ingredients using a 2SLS (Two-Stage Least Squares) approach with Instrumental Variables. In order to establish the causal relationship between generic drug entry and the result of the three questions mentioned above, exogeneity requirements must be satisfied in the generic entry decision. This requirement, widely used in econometrics, mandates that the choice of instrument variable should not correlate with any explanatory variables but only the generic entry decision. Paulo et al. (2013) take advantage of the objective nature of patents expiration and the government setting of auction dates to construct an instrument variable, which is further defined as the difference in days between patent expiration date and the tendering session starting date. By the two-stage regression with the inclusion of fixed effect on product, public body, and time, Paulo et al. (2013) suggest that the bidding price of branded suppliers is lowered in response to the entry of generic supplier and the price paid for pharmaceuticals reduces by seven percent due to the fierce competition by the new entry.

2.2.2 Pharmaceutical Tendering and Market Concentration

In addition to the bidding price change with pharmaceutical tendering, many studies have empirically identified how tenders have changed the market dynamics for pharmaceutical suppliers, including number of participants, market concentration index, and market dynamics.

Paulo et al. (2013) use a 2SLS model to examine the effect of a generic drug entry on the number of branded drug suppliers in the market. It is common knowledge that the fierce competition after the presence of a generic producer will discourage the participation of branded manufacturers. Paulo et al. (2013) validate this empirically with multiple versions of regression including General Least Squares, 2SLS, fixed effects with drug and buyer-specific features, time-specific fixed effects, and the health indicators of the municipalities where the manufacturer is based. The statistically significant negative parameter estimation on the dummy variable representing generic drug entry shows that the presence of a generic drug supplier indicates a reduction in the total number of suppliers for branded drugs in the market by 35 percent. Thus, the increased competition prevents one-third of branded suppliers from staying in the market.

While some researchers focus on quantifying market dynamics using the number of bidders in the market, other researchers such as Wouters et al. (2019) use the Herfindahl-Hirschman (HH Index) index to measure how fierce the market concentration level. Wouters et al. (2019) use the HH index, which not only considers the number of firms/suppliers in the market, but also considers the relative market shares of those firms. It is computed by summing up the squared market share for each supplier in the market. If all firms in the market have equal market shares, the HH index would be minimized. Thus, increasing the HH index indicates uneven distribution of market shares and it is scaled from 0 to 100000 with 0 denoting perfect competition. By calculating the HH indices over a 14-year period, Wouters et al. (2019) find that in general tendering does not change the overall moderate to competitive tendering market in Africa with HH index less than 2500. However, the authors also find that the number of firms actually winning the bid decreased in some drug categories. Overall, the tendering market in South Africa remains moderately competitive in most drug categories.

3 Tendering as Auction: Scope and Concept

Due to the complexity of real-world auction cases, in auction theory there are many different auction types, each of which denotes its own assumptions and requirements. In this section, we summarize the most common auction type for tendering and its assumptions. These assumptions correspond to a priori distribution of models for price estimation. Thus, finding the correct auction type is an essential step before fitting empirical models.

3.1 Tender Versus Auction

As introduced in Sect. 1, a tender is a closed price offer where each bidder keeps their own price, not knowing the other's price. Auctions are characterized as transactions with a specific set of rules detailing resource allocation according to participants' bids. They are categorized as games with incomplete information because in the vast majority of auctions, one party will possess information related to the transaction that the other party does not. In a general sense, tender can be viewed as a private form of auction where prices are not transparent to other parties during the bidding price. Thus, auction theory can also be used in the tender process to analyze optimal bidding strategies, find equilibrium bidding price as well as evaluate bidding design.

3.2 Independent Private Value Auction

The basic auction environment consists of the following four elements: the number of players (bidders) n, one object i (drug) to bid on, the actual value of the object v_i, and bidders signal s_i. The signal reflects how much each bidder values the bidding product i, which is not necessarily the same as the actual value of the bidding object. If each bidder has a different view about how much the bidding product means to them, namely each bidder has a different private information (signal) and if knowing the other's signal may change the perception of their own value, this would be a common value model. In contrast, during the tendering process, each bidder has their own private information about the value of the bidding product. This signal would not change even if information from others is required. This is called the Independent Private Value (IPV) auction. Under the normal tendering setting, tendering is typically an IPV auction. Using the notations, this means, bidder i's information (signal) is independent of bidder j's information. Moreover, bidder i's value is independent of bidder j's information—so bidder j's information is private in the sense that it does not affect anyone else's valuation.

3.3 First Price Sealed Auction

IPV auction focuses on whether the information and value for each bidder is independent. In addition to independency, another very important element in auctions is winning prices. There are multiple forms of auction such as absolute auction, minimum auction, sealed auction, reserve auction, etc., depending on the winning rule and completeness of information on other bidders. In a sealed bid auction, bidders submit their bid $b_i, \ldots b_n$ simultaneously, and the bidder who offers the highest bid (in tender case, lowest tendering price) wins the bid and pays the bidding prices offered. "Sealed" means each bidder submits the bid privately so no one else knows the information, while "First Price" means the winning bidder pays the winning price, not the second winning price. It is quite obvious that bidders would not bid on their true values because this brings no profit/benefit. Normally by bidding a little bit lower (in tender cases higher tendering price), bidders can gain margins.

3.4 Number of Bidders

Number of bidders in a single bidding round reflects the scale of a specific bidding process. Number of bidders highly depends on the number of competing manufacturers in the market. For biosimilar bidding, the number of bidders is usually small while for other generic drugs, the number of bidders could be quite large. Yao and Tanaka (2015) empirically examine the determinants of bidding behavior by using the provincial-level data consisting of 2758 bidders, with 757 being the highest number of manufacturers in the same bidding round. So far, this is a truly large-scale study with observational data. For literature incorporating structural models, researchers tend to use a finite number of bidders in the simulation stage and quite a small number of bidders in the empirical estimation. For example, Yao and Tanaka (2015) consider $n = 7$ bidders in the Monte Carlo Simulation and during the empirical illustration, the paper used California Highway Procurement data with 2–7 bidders. Guerre et al. (2000) used $n = 5$ bidders in the classical paper proposing non-parametric estimation of first-price auctions. To sum up, for empirical studies using structural models or non-parametric models, the number of bidders normally ranges from 2 to 10. For studies using observational studies, the number of bidders could be much greater.

4 Empirical Methods for Price Auction Estimation

The previous three sections summarized the pharmaceutical tendering background and auction types qualitatively. In this section, we will focus on the methodology to estimate the most important parameter in the tendering price—bidding price. As introduced in Sect. 3, most tendering processes could be modeled theoretically by auction theory. In empirical literature, first-price auctions are the auction type that has been researched most frequently, through structural or reduced form methods. Structural approaches mainly focus on recovering the distribution of observed bids or bidder's private value in order to extrapolate the unobserved valuation of the bidders. Considering the nature of all structural estimations, they rely heavily on assumptions of the underlying distribution, which is indirectly reflected by the auction type in terms of information asymmetry, information completeness, and the presence of signaling. Reduced form estimations mainly target at solving the selection bias problem, where endogenous factors can bias the estimation results of exogenous variables. By rearranging the equations algebraically until every endogenous variable is on the left side of the equation and all the exogenous variable (also including lagged endogenous ones) is on the right side of the equation so that potential selection bias has been resolved. In this section, we discuss these two methods through the estimation on bidding price.

4.1 Bidding Price Determinants Estimation with Reduced Form Approach

Yao and Tanaka (2015) empirically examine the determinants of suppliers' bidding behavior in a multiple bidding setting using a provincial pharmaceutical procurement dataset consisting of 19818 biddings from 2758 bidders on 37 groups of drugs from 2007 to 2009. Considering the bidding setting where the winning price is only known to the final winner, bidding price would introduce selection problems when using a non-random subsample to use that price to all participants. Thus a simple OLS or GLS regression with bidding price as the dependent variable is not feasible. In presence of selection bias, the paper used the Heckman Selection Model to correct for endogeneity. Yao and Tanaka (2015) start with the simplest model construction:

$$y_{ijt} = \mathbf{x}_{ijt}\beta + u_{ijt} \tag{1}$$

$$s_{ijt} = \mathbf{1}\left\{s^*_{ijt} > 0\right\} \tag{2}$$

$$s^*_{ijt} = \mathbf{z}_{ijt}\gamma + v_{ijt} \tag{3}$$

where y_{ijt} is the bidding price for bidder j on product i in year t, x_{ijt} are vectors of explanatory variables that would affect bidding prices. The paper includes number of bidders, spatial distance between the bidder and the auctioneer, product specific features including product range and standards, quality specific features including the technological skills, experience, and multiple bidding potential. In Yao and Tanaka (2015)'s definition, u_{ijt} represents the idiosyncratic error which is not correlated to any of the variables and v_{ijt} represents the idiosyncratic error during selection. s_{ijt} is a latent variable and Eqs. (2) and (3) jointly describe the choice made by the auctioneer. Directly estimating the above two equations or using maximum likelihood alternatively both suffer from multicollinearity. The authors used the panel data version of the Heckman selection model with the Mundlak-Chamberlain approach to correct the selection bias as follows:

$$y_{ijt} = \mathbf{x}_{ijt}\beta + c_{ij} + u_{ijt} \tag{4}$$

$$\tilde{s}_{ijt} = \mathbf{1}\left\{\eta_t + \mathbf{z}_{ijt}\gamma_t + \mathbf{z}_{ij}\varphi_t + e_{ijt} > 0\right\} \tag{5}$$

$$e_{ijt} = a_{ij} + v_{ijt} \tag{6}$$

where c_{ij} is the unobservable and z_{ij} is a vector representing the means of x_{ijt}. Since e_{ijt} is not depending on z_{ij}, Eq. (1) is converted to the following equation:

$$y_{ijt} = \mathbf{x}_{ijt}\beta + c_{ij} + \rho E\left(e_{ijt} \mid \mathbf{z}_{ij}, \tilde{s}_{ijt}\right) + v_{ijt} \tag{7}$$

By first estimating the possibility of selection, pooled OLS is then applied to the selected sample and the paper discovers that more fierce competition and more winning experience encourage suppliers to bid much more aggressively on the bidding prices and consequently make lower bids. In addition, those bidders that are in less competitive groups are less sensitive to the number of participants and their past winning times.

4.2 Structural Estimation of Auction Models

The empirical nature that uses structural econometric modeling approaches to understand firm and consumer behavior has attracted wide interest. The estimation of auction mainly focuses on the private value of the bidders. The earliest literature on structural estimation of auction prices dates back to 1992, where Paarsch (1992) introduce parametric structural models to estimate private and common value first-price sealed auctions. After that, much literature examine the same topic using parametric structural models, including (Donald and Paarsch 1993; Elyakime et al. 1994; Flambard and Perrigne 2022; Campo 2022). All of these papers model the game as a first-price auction game where the Bayes-Nash equilibrium is the end point and the idea to parametrically estimate the distribution of bidder's private value is achieved by inverting the equilibrium function of private value from observed bids. Guerre et al. (2000) introduce the two-stage optimal estimation of private value in a non-parametric way. This non-parametric method has been widely researched and applied in following studies, with the emergence of another approach, called the quantile-based estimation. In this subsection, we'll first go over the structural literature of first-price auctions and the experimental evidence of the estimations. Then, we'll discuss the non-parametric development of structural estimation, including two-stage estimation and quantile-based estimation.

4.2.1 Reasonable Structural Estimation of Price Auction

Bajari and Hortacsu (2003) use experimental data to examine whether four different kinds of structural models give reasonable estimates on bidders' private value. Many researchers have challenged the strict rationality assumptions imposed by parametric models to be infeasible in reality. The critics focus on mapping the estimation results, bidders' valuations to the bidders' true private information. Bajari and Hortacsu (2003) structurally estimated four models under the first-price auction setting: risk neutral Bayes-Nash, risk averse Bayes-Nash, Quantal Response Equilibrium (logit equilibrium model), and an adaptive model of learning.

Assuming there are $i = 1 \ldots N$ symmetric bidders in the market and v_i denoting the valuation of one single and indivisible product. Each bidder's valuation v_i is iid with cdf $F(v)$ and pdf $f(v)$; b_i denoting the simultaneously submitted bid by bidder i; u_i denoting the utility.

Then under the risk neutral Bayes-Nash model, the first-order condition for maximizing expected profit can be expressed by

$$v = b + \frac{F(\phi(b))}{f(\phi(b))\phi'(b)(N-1)} \quad (8)$$

Using $G(b)$ and $g(b)$ as the distribution and density of the bids, the above equation can be written as follows:

$$v = b + \frac{G(b)}{g(b)(N-1)} \quad (9)$$

Next under the risk averse Bayes-Nash model, the first-order condition is

$$v_i = b_i + \frac{\theta_i}{Y_i(b_i)} \quad (10)$$

where $1-\theta_i$ represents Bidder i's risk preference, also known as the coefficient of relative risk aversion.

Thirdly, under the logit equilibrium model, with the extreme value distribution:

$$\sigma(b_i; v_i, \mathbf{B}) = \frac{\exp(\lambda \pi(b_i; v_i, \mathbf{B}))}{\sum_{b' \in \mathcal{B}} \exp(\lambda \pi(b'; v_i, \mathbf{B}))} \quad (11)$$

where $\mathbf{B}(b \mid v)$ represents a symmetric strategy, which is a measure that gives every bid b a probability based on the condition upon a valuation v. An equilibrium is a bidding function $\mathbf{B}(b \mid v)$ that is a fixed point, that is $\mathbf{B}(b \mid v_i) = \sigma(b_i; v_i, \mathbf{B})$.

Finally, for the simple adaptive model, the first-order condition for maximizing expected profit is

$$\hat{v}_{it} = b_{it} + \frac{\hat{G}(b_{it} \mid h_{it})}{\hat{g}(b_{it} \mid h_{it})(N-1)} \quad (12)$$

By using the structural estimation to measure the closeness of the estimated valuation distribution of the above four models with the true distribution of bidders' private value, Bajari and Hortacsu (2003) calculate the Kolmogorov-Smirnov distance between the distribution and true values. The paper notices that given the number of bidders is large enough, the three models except QRE are able to uncover the deep parameter. When the number of bidder is small, the estimation result is not stable enough and is more sensitive to the choice of models. Rational models give better estimated results than the behavioral models.

4.2.2 Two-Stage Non-parametric Structural Estimation of Price Auction

Guerre et al. (2000) point out that the structural econometrics approach on price auction has to rely exclusively on the parametric requirement of the distribution of bidders' private value. However, the strong assumption may not be met completely in both observational and experimental data. Guerre et al. (2000) also mention the computational limitation for structural models due to its dependency on complex numerical computations and simulations to find out the Bayesian Nash equilibrium. Instead, Guerre et al. (2000) propose a two-stage indirect process to estimate the distribution of bidder's valuation from observed bids without computing Bayesian Equilibrium or imposing any parametric assumptions on the observed bids.

The main idea of Guerre et al. (2000)s two-stage approach is that constructing a function with the distribution of observed bids, corresponding bids, and corresponding density function to represent the private value of each bid. This idea is implemented in the first step by making a set of pseudo-private values according to the observed bids' kernel distribution and density function. Then, the density of the bidder's private value can be estimated non-parametrically using the pseudo samples. Guerre et al. (2000) prove the uniform consistency property of the estimator by showing it has the best uniform convergence rate in estimating the latent density.

For a more detailed model specification, let i denote bidder $i = 1 \ldots I$ and v_i as the private value for the bidding product, and p_0 as the reservation (lowest possible) price. Let $F(v_i)$ denote the common distribution of private values and $f(v_i)$ denote the density. Under the Bayesian Nash Equilibrium of symmetric bidders, the equilibrium bid b_i for bidder i is:

$$1 = (v_i - s(v_i))(I - 1)\frac{f(v_i)}{F(v_i)}\frac{1}{s'(v_i)} \tag{13}$$

With non-parametric identification, the previous equation can be rewritten that now expresses the individual private value vi as a function of the individual's equilibrium bid b_i, its distribution $G(\cdot)$, its density $g(\cdot)$ as follows:

$$v_i = \xi(b_i, G, I) \equiv b_i + \frac{1}{I-1}\frac{G(b_i)}{g(b_i)} \tag{14}$$

For estimation, considering L auctions, $G(\cdot)$ and $g(\cdot)$ can be estimated by

$$\tilde{G}(b) = \frac{1}{IL}\sum_{l=1}^{L}\sum_{p=1}^{I} 1\left(B_{pl} \leq b\right) \tag{15}$$

$$\tilde{g}(b) = \frac{1}{ILh_g}\sum_{l=1}^{L}\sum_{p=1}^{I} K_g\left(\frac{b - B_{pl}}{h_g}\right) \tag{16}$$

where h_ρ is a bandwidth and $K_g(\cdot)$ is a kernel with a compact support. In this paper, the author uses a tri-weight kernel and uses 1.06 as the bandwidth. We'll discuss more about the choice of kernel and bandwidth in the last section of this review with real data.

Then as an illustration, the paper conducts a Monte Carlo Simulation with 200 auctions and 5 bidders over 1000 observed bids to empirically verify that the latent density of bidder's private values can be estimated from available bids.

4.3 Quantile-Based Non-parametric Estimation of Private Value

After Guerre et al. (2000)s paper on the two-stage optimal non-parametric structural estimation approach, numerous studies have focused on releasing the assumption on the distribution of observed bids. Marmer and Shneyerov (2012) identify the idea to estimate a bidder's valuation distribution in another flavor, estimating it based on the quantile representation of the first-order condition. Under the risk neutral Bayes-Nash model, the Guerre et al. (2000) paper transform the first-order condition for optimal bids and expresses a bidder's value as an explicit function of the submitted bid, the Probability Density Function (PDF) and Cumulative Distribution Function (CDF) of bids, as shown in Eq. (10). Marmer and Shneyerov (2012) further propose to estimate the valuation distribution based on the quantile representation of the first-order condition. This indicated, when there is the strict monotone condition on the equilibrium bidding strategy, the valuation quantile function $Q_v(\cdot)$ can be expressed as

$$Q_v(\alpha) = Q_b(\alpha) + \frac{1}{I-1} \frac{\alpha}{g(Q_b(\alpha))'} \quad 0 \leq \alpha \leq 1 \qquad (17)$$

where $Q_b(\cdot)$ is the bid quantile function. In Eq. (17), Marmer and Shneyerov (2012) propose to first estimate $Q_v(\cdot)$ using plug-in estimators for $g(\cdot)$ and $Q_b(\cdot)$ and subsequently estimate the valuation density using the relationship $f(v) = 1/Q'_v\left(Q_v^{-1}(v)\right)$. They prove that the quantile-based estimator is asymptotically normal and has the optimal rate of Guerre et al. (2000)s definition of Private Value (GPV).

Marmer et al. (2010) also use the quantile-based non-parametric approach to infer the PDF of private values on first-price auctions under the Independent Private Value (IPV) construction. They disclose a fully kernel-based estimator of the quantiles and PDF over observed bids and estimate it non-parametrically. They are also able to achieve the optimal rate with Guerre's paper and under proper choice of bandwidth, the estimator is also asymptotically normal. While Marmer et al. (2010)s paper have already released one of the steps in Guerre et al. (2000)s paper on constructing pseudo values, Luo and Wan (2017) go one step further and propose a fully tuning-parameter-free estimator for the valuation quantile function. This means to estimate the quantile function of the valuation requires neither the choice of a kernel nor a bandwidth. They provide a trimming-free smoothing estimator and this estimator is also asymptotically normal and is the same as the optimal rate of Marmer et al. (2010)s paper. Liu and Luo (2017) investigate the comparison of valuations in first-price auctions using non-parametric tests. Building on the fact that two distributions of private valuations would be the same if and only if the integrated quantile functions are identical, the paper proposed a test statistic that measures the square distance between the sample analogues of the linear functional for bid samples.

5 Non-parametric Estimation on Observational Data

In this section, we briefly introduce the process of applying non-parametric structural estimation on real-world observational tendering data. This includes a description of a sample dataset coming from one of the affiliates of a large pharmaceutical company, the initial findings from basic data visualization, the process to formulate the empirical model, the choice of parameters during the non-parametric estimation, and the conclusions from the estimation.

5.1 Dataset

The dataset comes from one of the affiliates of a large multi-national pharmaceutical company, containing around 1500 tendering records over a three-year period from 2018 to 2020. Tender records cover three main generic products in oncology with biosimilar competition. Tendering process in the concerned country follows the typical first-price sealed bid auction where competitors place the tender offer simultaneously without knowing the other's price as signals. In addition, through market research and discussion, for the three products involved in the tendering process, we can assume there are three main participants in the market (including the manufacturer). Figure 1 gives the description of the dataset:

5.2 Visualization and Descriptive Analysis

With real-world tendering result datasets, based on our experience, we first start by visualizing and cleaning the dataset with a few plots and descriptive statistics. We will explain in more detail which variables and information have been used to perform exploratory data analysis and what information could be relevant to down-streaming modeling tasks. The first dimension to explore is the tender-specific information and we plot basic descriptive statistics like bar plot density plot on categorical variables such as Tender Type, Hospital (Customer), Hospital Type, Tender Result, Bidding Product, Discount Rate, and kernel density plot on numerical variables such as Manufacturing Price, Winning Price, and Wholesale Price Margin. This information gives

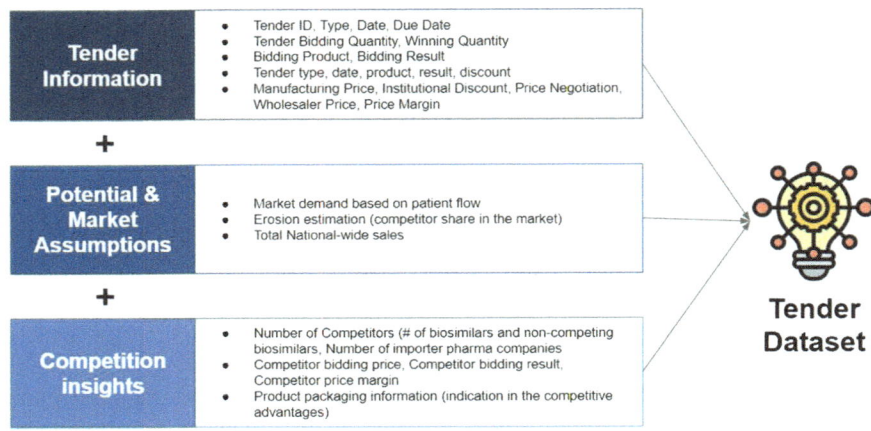

Fig. 1 Three dimensions of tender dataset

us an overview of the distribution of tender-related attributes. Similar to our assumptions to real-world bidding setting, most of the categorical are skew-distributed with highly imbalanced classes. We account for that in our estimation by using re-sampling methods to create more balanced classes by imposing different class weights.

In addition to visualizing variables in single dimensions, we also perform pivot analysis across multiple dimensions, such as bar plot of tender results by product type, bar plot of tender type by product, bar plot of tender result by tender type, and bar plot of tender result by tender quantity (convert numerical quantity to categorical intervals). For confidentiality reason, we only describe the methodologies in visualization, rather than the actual results. These plots provide us with the interactions between variables and help us to verify some of the business assumptions, which will be further explained in more detail in the next section. To summarize, exploratory data analysis and descriptive statistics are a critical step to perform on a real-world dataset as the starting point before modeling and estimation. It provides initial insights and helps validate some of the key assumptions to choose the most accurate auction forms.

5.3 Assumptions Validation

As described in Sect. 3, Tendering Scope and Auction Forms are defined by a set of conditions and assumptions. Choosing the most accurate auction form is the most critical step for modeling pharmaceutical tendering in a real-world setting. However, in most real-world cases, the datasets do not completely follow one exact type of auction form and some of the assumptions need to be verified for modeling. In this part, we'll illustrate how to perform validity checks on real-world datasets for the four most important assumptions to determine the corresponding auction form.

First, from an auction theory perspective, the starting point is to examine the number of players in the game, which maps the number of manufacturers (bidders) in a pharmaceutical tendering setting. This information is examined by a simple frequency count of the Manufacturer (bidder) in the dataset. One important check is to confirm that the products from manufacturers are perceived as "Biosimilars" to each other with the same standards of quality, safety, and efficacy since first-sealed auction form requires each bidder to bid on non-differentiated products. For non-competing biosimilar manufacturers, which offer low quality products or products with different biologic compositions are not eligible to be counted as players (bidders) in the modeling process. Thus the key point to check is the count of manufacturers in the dataset and keep those who offer biosimilar products.

Next, the validity check should focus on the bidding rule. This includes both qualitative and quantitative checks. Qualitative check deals with the following questions relevant to tender managers or tender specialists who manage and perform the actual bidding process: What is the tender submission process? Does the participating entity have any information about the other competitors before the bidding process? Are bidding prices submitted simultaneously in each round? Or sequentially where some competitors could signal the others' behavior? How many rounds of bidding are contained in a full bidding process? Is there any entry threshold for each round and how is it defined? These questions require qualitative checks with business stakeholders or tender managers who actually participate in the tendering process. In addition, the number of rounds in a tender process can be quantitatively verified in the dataset by grouping records by unique tender id (if it exists) and count the number of records under each tender id.

Then, the most important validity check is on the winning rule since first-price sealed bid auction requires "lowest price winning rule" which means the winner is solely determined by the raised bidding price, with no other factor taken into account such as brand perception and loyalty, quality, and packaging. This assumption check is performed quantitatively on the dataset by first grouping all the bidding records by the unique identifier of a tender, for example, tender id (if it exists), and extracting the lowest bidding price under the same id; next, extract the final winning price of each tender and comparing it with the lowest bidding price to check whether the two figures are equivalent. Theoretically, with the "lowest price winning rule" the lowest bidding price in a tender should be the same as the final winning price of the bid. However, in real-world bidding datasets, sometimes the two figures can be different. The underlying reason could be that the manufacturer who offered the lowest bidding price could not fulfill the bidding quantity completely due to logistics blocker or insufficient remaining quota. Under this situation, those tender results should be excluded from the modeling to avoid adding noise to the models and estimation. Thus it is very crucial to perform a quantitative check to validate the "lowest price winning rule".

Finally, based on our analysis on the dataset, verifying the winning quota is the last step for the validity check. This includes would the winner take all the quota as submitted in the bid (Winners take all)? Or the ending quota could be smaller/larger than the submitted amount? In some cases, the winner cannot get exactly the same amount as submitted during the bidding price. This happens most frequently when there is a mix of biosimilar manufacturers and non-competing biosimilar manufacturers in the bidding round. And in the above case, there is a very high probability that the previous "lowest price winning rule" does not hold. Winners may only get a subset or part of the submitted bidding quantity and the remaining small portion may go to the non-competing biosimilars due to their extremely low price compared to biosimilars. This assumption on quota is generally examined in the dataset by comparing the column indicating submitted quota and the column indicating winning quota. If these two columns are not the same, normally we should dive deeper to check which one is larger and whether the pattern is consistent. At the same time, this assumption is always examined together with the previous one "winning rule". If the winning quota diverges with the submitted quota, we must be very cautious on choosing the auction form since the first-price sealed bid auction does not apply anymore.

5.4 Modeling and Estimation

After checking the assumptions on the real-world tendering dataset, we confirm that for the country concerned, the bidding form can be well modeled as a first-price sealed bid auction. After data cleaning and processing, we follow the methodology proposed by Guerre et al. (2000) with Two-Stage Non-Parametric Structural Estimation to compute the Cumulative Density Function of the private value for bidders.

Using the Two-Stage Non-Parametric approach by Guerre et al. (2000), we extract the Wholesaler Price after Discount (MTS price) in the dataset to obtain the observed bids and use $I = 3$ to represent the three main suppliers in the market. Following the two-stage estimation process, first according to Eq. (14), we are able to construct a sample of pseudo-private values using non-parametric estimates of the distribution and density functions of observed bids (MTS Price). Then for the second step, using Eqs. (15) and (16), we are able to get the distribution of the bidders' valuation with the choice of bandwidth and kernel. Similar to Guerre et al. (2000)s method, we also try the Triweight Kernel and the Epanechnikov Kernel. For bandwidth, we use a loop to select the most reasonable bandwidth ranging from 0.35 to 2.6 (the above range is decided by observing the pattern of the distribution plot). After obtaining the distribution of valuation, we use KS test (Kolmogorov-Smirnov test) and standardization to map back to the bidding price of actual bids in order to get the Probability Density Function and Cumulative Distribution Function for the probability to win. After obtaining the Cumulative Distribution Function of bidder's private values, given a constant bidding price x, with the definition of Cumulative Distribution Function, we can easily get the probability that the X will take a value less than or equal to x. According to the bidding setting, if the bidding price of other competitors is less than the bidder's offering price, the bidder will lose this bid. This means the area under the Cumulative Distribution Function curve and to the left of the raised price x denotes the probability that the competitor price will be lower than the raised bid. Thus, the winning probability is obtained by subtracting the probability from 1.

5.5 Adjustments and Lessons Learned

In the previous section, we discussed how to perform validity checks on the real-world tendering data to determine the most appropriate auction forms. This is the most important step since in real-world settings, it is hard to meet all the assumptions of a specific auction form due to the complexity of bidding system design and actual bidding process implementation. At the same time, policy and regulatory requirements also impose some constraints on meeting all the requirements and assumptions of an auction type. This directly leads to the fact that during the modeling and estimation process, we often need to use "rule of thumb" to determine some of the parameters that best fit and depict the dataset, instead of pre-defined parameters from previous assumptions. One example in our estimation case is the choice of Kernels and their bandwidth using the two-stage methodology in the Guerre et al. (2000) paper.

All the estimation process in the real-world tendering data is implemented in Python 3.6. When selecting the kernel, there is no clear evidence that the tri-weight kernel should outperform other kernels. Also for the bandwidth choice, it is more of a rule of thumb instead of mathematically proved constant. As explained in the previous paragraph, finding the best combination of kernel and bandwidth with a real-world tendering dataset requires iterative computation and it is a more heuristic process. Thus, we tried all the possible combinations of bandwidth and kernel to get the most meaningful valuation distribution.

In terms of take-aways and lessons learned from fitting observational data, we find that a well-established structural model is the theoretical foundation. In addition to that, adjustments should also be made to understand the data and give intuitive explanations to the data. Different choice of parameters results in different distributions of valuations, where the probability to win should not have multiple spikes in real-world bidding practices. Thus, fitting non-parametric structural models should always consider the underlying business rationale.

References

Bajari, P., & Hortacsu, A. (2003). Are structural estimates of auction models reasonable? *Evidence from Experimental Data*. https://doi.org/10.3386/w9889.

Campo, S. (2022). Asymmetry and risk aversion within the independent private paradigm: The case of the construction procurements, UNC Working Paper.

Donald, S. G., & Paarsch, H. J. (1993). Piecewise pseudo-maximum likelihood estimation in empirical models of auctions. *International Economic Review, 34*(1), 121. https://doi.org/10.2307/2526953.

Elyakime, B., Laffont, J. J., Loisel, P., & Vuong, Q. (1994). First-price sealed-bid auctions with secret reservation prices. *Annales D'Économie Et De Statistique,34*, 115. https://doi.org/10.2307/20075949

Flambard, V., & Perrigne, I. (2022). Asymmetry in procurement auctions: Some evidence from snow removal contracts, unpublished working paper, University of Southern California.

Gray, A., & Suleman, F. (2014). Pharmaceutical pricing in South Africa. *Pharmaceutical Prices in the 21st Century*, 251–265. https://doi.org/10.1007/978-3-319-12169-7-14

Guerre, E., Perrigne, I., & Vuong, Q. (2000). Optimal nonparametric estimation of first-price auctions. *Econometrica, 68*(3), 525–574. https://doi.org/10.1111/1468-0262.00123.

Liu, N., & Luo, Y. (2017). A nonparametric test for comparing valuation distributions in first-price auctions. *International Economic Review, 58*(3), 857–888. https://doi.org/10.1111/iere.12238.

Luo, Y., & Wan, Y. (2017). Integrated-quantile-based estimation for first-price auction models. *Journal of Business & Economic Statistics, 36*(1), 173–180. https://doi.org/10.1080/07350015.2016.1166119.

Maniadakis, N., Holtorf, A., Corrêa, J. O., Gialama, F., & Wijaya, K. (2018). Shaping pharmaceutical tenders for effectiveness and sustainability in countries with expanding healthcare coverage. *Applied Health Economics and Health Policy, 16*(5), 591–607. https://doi.org/10.1007/s40258-018-0405-7

Marmer, V., Shneyerov, A. A., & Xu, P. (2010). What model for entry in first-price auctions? A nonparametric approach. *SSRN Electronic Journal*. https://doi.org/10.2139/ssrn.1628367

Marmer, V., & Shneyerov, A. (2012). Quantile-based nonparametric inference for first-price auctions. *Journal of Econometrics, 167*(2), 345–357. https://doi.org/10.1016/j.jeconom.2011.09.020.

Paarsch, H. J. (1992). Deciding between the common and private value paradigms in empirical models of auctions. *Journal of Econometrics, 51*(1–2), 191–215. https://doi.org/10.1016/0304-4076(92)90035-p.

Paulo, A., Kleinio, B., & Dante, G (2013). Generic-branded drug competition and the price for pharmaceuticals in procurement auctions. In *Working Paper*

Petrou, P. (2016). Long-term effect of tendering on prices of branded pharmaceutical products. *Health Policy and Technology, 5*(1), 40–46. https://doi.org/10.1016/j.hlpt.2015.10.006.

Petrou, P., & Talias, M. A. (2014). Tendering for pharmaceuticals as a reimbursement tool in the cyprus public health sector. *Health Policy and Technology, 3*(3), 167–175. https://doi.org/10.1016/j.hlpt.2014.04.003.

Petrou, P., & Talias, M. A. (2015). Price determinants of the tendering process for pharmaceuticals in the cyprus market. *Value in Health Regional Issues, 7*, 67–73. https://doi.org/10.1016/j.vhri.2015.09.001.

Simoens, S., & Cheung, R. (2019). Tendering and biosimilars: What role for value-added services? *Journal of Market Access & Health Policy, 8*(1), 1705120. https://doi.org/10.1080/20016689.2019.1705120.

Wouters, O. J., Sandberg, D. M., Pillay, A., & Kanavos, P. G. (2019). The impact of pharmaceutical tendering on prices and market concentration in South Africa over a 14-year period. *Social Science &; Medicine, 220*, 362–370. https://doi.org/10.1016/j.socscimed.2018.11.029

Yao, Y., & Tanaka, M. (2015). Price offers of pharmaceutical procurement in China: Evidence from Guangdong province. *The European Journal of Health Economics, 17*(5), 563–575. https://doi.org/10.1007/s10198-015-0700-2.

Philipp Mekler Trained in both biochemistry (Ph.D.) and mathematics (M.Sc.), with more than forty years of experience in the Pharma/Life Science sector in R&D, sales/marketing, business analysis and bio-business finance/venture capital (in CH, US, & IL). Currently at Roche Pharma International as Strategic Advisor for a Data & Analytics unit.

Jingshu Sun Data Scientist in the Pharma International Informatics Data and Analytics Chapter at Roche, where she works on a wide range of analytics topics on competition, pricing and sales channel. Methodology wise, she combines econometrics, economics models with machine learning approaches to draw insights from both structured and unstructured data sources.

Open Access This chapter is licensed under the terms of the Creative Commons Attribution 4.0 International License (http://creativecommons.org/licenses/by/4.0/), which permits use, sharing, adaptation, distribution and reproduction in any medium or format, as long as you give appropriate credit to the original author(s) and the source, provide a link to the Creative Commons license and indicate if changes were made.

The images or other third party material in this chapter are included in the chapter's Creative Commons license, unless indicated otherwise in a credit line to the material. If material is not included in the chapter's Creative Commons license and your intended use is not permitted by statutory regulation or exceeds the permitted use, you will need to obtain permission directly from the copyright holder.

Multi-Echelon Inventory Optimization Using Deep Reinforcement Learning

Patric Hammler, Nicolas Riesterer, Gang Mu, and Torsten Braun

1 Introduction

The operation of supply chains is a major cost driver for all manufacturing companies. It is imperative to keep this cost at a minimum and the service level at a maximum to enable companies to redirect investment to their core goals, such as the development of new drugs in the healthcare industry. The field that deals with this task is called inventory management and has served as an intensely studied research area for many decades. In practice, companies often rely on parameterized reorder policies for the operation of inventory management (e.g., De Kok et al. 2018). These consist, for example, of a periodic reorder timing (T) and a reorder quantity (Q) that depends on the difference of the current inventory on hand (IOH) and the target IOH. The conceptual designs of such parameterized reorder policies are usually hand-crafted and based on historical experiences, sales forecast information, and safety stock considerations. Parameterized reorder policies are intuitive and easily applicable—on the other hand, they tend to be an oversimplified solution for a complex challenge due to the stochastic characteristics of the problem: E.g., demand anomalies require a situational T and Q, which highlights the importance of so-called dynamic reorder policies.

P. Hammler (✉) · T. Braun
Universität Bern, Bern, Switzerland
e-mail: patric.hammler@inf.unibe.ch

T. Braun
e-mail: braun@iam.unibe.ch

N. Riesterer
F. Hoffmann-La Roche AG, Basel, Switzerland
e-mail: nicolas.riesterer@roche.com

G. Mu
University of Zurich, Zürich, Switzerland
e-mail: gang.mu@math.uzh.ch

Finding an optimized, dynamic reorder policy for a given network of inventory systems is a challenging task. With recent advances in the field of Artificial Intelligence (AI), the question naturally arises: Can AI help to make better decisions, and thus, reduce the cost for the operation of supply chains? This question is justified, especially when one considers that the best chess player in the world is not a human being anymore (Silver et al. 2017). Just like chess, inventory management is a challenge in which the optimal sequence of decisions is sought. If with chess, we are looking for the optimal sequence of decisions that maximize the chances of winning, in the case of inventory management we are looking for the optimal sequence of decisions that minimize the cost.

Reinforcement Learning (RL) is the paradigm in the field of machine learning dedicated to learning an optimized policy in sequential decision-making challenges. At a high level, an agent takes situational decisions and receives feedback on the quality of the agent's decision in return. As a consequence, the agent takes the feedback into account to improve the policy. In the latest research publications, policies are based on deep neural networks, in which case the methodology is referred to as Deep Reinforcement Learning (DRL). DRL has recently attracted considerable attention: An RL agent beats professional players in the classic board game Go, which is considered to be the most challenging board game use-case for AI due to its high number of possible combinations (Silver et al. 2016). DRL enables autonomous driving (Kiran et al. 2020) and can potentially be leveraged for precision dosing in the healthcare sector (Ribba et al. 2020). In this chapter we review the applicability of DRL for Multi-Echelon Inventory Optimization (MEIO).

This chapter aims to provide an introduction to the domain of MEIO with RL and is structured into eight sections: Sect. 2 introduces the term MEIO and explains the multiple challenges that are connected to it from an optimization perspective. Section 3 provides a brief overview of research streams in the field MEIO. Section 4 connects the topics MEIO and RL before Sect. 5 provides a detailed introduction to the concept of RL. Section 6 showcases an experiment to evaluate the applicability of RL in MEIO challenges. In Sect. 7 the results are discussed. Section 8 provides an outlook of potential future research streams. Section 9 showcases conclusions and provides an outlook of future research efforts.

2 Challenges of Multi-Echelon Inventory Management from an Optimization Perspective

The goal of inventory management optimization is to optimize the reorder policy in a way so that the cost related to the operation of inventory systems are minimized. This section introduces five layers of complexity explaining why inventory management is a highly challenging optimization task:

The major operational cost of inventory systems can be structured into holding-, shortage-, and reordering costs. This aspect introduces the first layer of complexity with regards to the optimization challenge: Minimizing holding cost through lowering the inventory level increases the likelihood of a stock-out and the associated shortage costs. To prevent stock-out, the ordering frequency can be increased, whereby this impacts the reordering cost on the other hand. This observation suggests that each cost category is interrelated leading to a complex, non-linear cost function.

The second layer of complexity is due to the stochastic characteristics of an inventory system: Each day, the IOH is reduced by the number of outgoing items (e.g., because of sales). This value depends on customer behavior, which can be estimated but is always associated with uncertainties, which is why this challenge falls under the category of stochastic optimization. Next to the demand, the lead time, which is defined as the time duration between order placement and supply delivery is another stochastic parameter. As a consequence, the decision-making must be optimal under the consideration of uncertainty.

The third layer is due to the interconnected characteristic of reordering policies within a supply chain distribution network. This can be briefly illustrated by an example: A large reorder from one warehouse can use up the entire reserves of the parent warehouse, with all further reorders from other warehouses subsequently no longer being able to be serviced. This example makes clear why the optimization must be carried out holistically and not in a warehouse-by-warehouse manner. Optimization approaches addressing this holistic problem characteristic are referred to as MEIO.

The fourth layer results from the third layer: Given the fact that the optimal ordering policy of an individual inventory system can only be found if the reorder policy of the entire inventory system network is optimal, this leads to a high number of parameters that have to be optimized. Many algorithms that still meet the requirements related to a highly complex cost function, stochastic system dynamics, and holistic optimization fail to scale to real-world supply chain dimensions. The holistic view of the problem leads to a dilemma: Either one resorts to high-performance algorithms, which outperform the rule-based approaches by far, or one wants to perform a holistic optimization, in which case the high-performance algorithms are not applicable anymore. This is probably also the reason why many companies still use comparably simple methods such as rule-based control.

The fifth layer is the variability of model and optimization goal assumptions: Multi-Echelon Inventory Systems (MEIS) can be divergent, convergent, sequential, or mixed (Clark and Scarf 1960). The policy to be optimized can either be periodic (all reorders at fixed time intervals) or dynamic (De Kok et al. 2018). The same applies to the reorder quantity: This can be flexible for certain applications—in other cases, the lot sizes are fixed and the optimization goal is to select the best available option. It can be seen that optimization algorithms must be adapted to the specific situation and that the underlying model assumptions are highly variable. In the next section, it will be shown that many traditional optimization approaches have to be tailored exactly to the problem—and general applicability is not given.

In the remainder of this section, we want to emphasize a potential solution to this dilemma—before that we want to have a deeper look at existing research efforts in the MEIO domain.

3 Literature Review of Inventory Management

Due to the immense relevance and high complexity of multi-echelon inventory management, many research paths have developed in the area with the first major research papers dating back to the 1960s. De Kok et al. (2018) composed a comprehensive literature overview on stochastic multi-echelon inventory models. In fact, most of the research efforts from the early days focused on the development of exact models: In Clark and Scarf (1960) a mathematical proof was presented that the reorder policy of an individual warehouse can only be optimal if the reorder policy is optimal for the entire network. However, due to the complexity of the problem, these models are based on highly-simplified assumptions limiting the applicability to real-world supply chains (Gijsbrechts et al. 2021). De Kok et al. (2018) state that developing optimal policy structures has turned out to be intractable. This fact, combined with the technological development in the semiconductor field and the associated increase in computational capacities, has led the research focus shifting to other methodologies such as parameterized, simulation-based, and approximative policy optimization. These are by no means completely separate fields of research—many seminal papers proposed a combination of the aforementioned algorithm categories. In order to provide the reader with an easy-to-understand intuition, the three areas are discussed separately below.

Parameterized policy optimization experienced its rise in the 1990s (De Kok et al. 2018). One prominent representative of parameterized policies is the base-stock policy, also known as (s, S) inventory control policy. Each time when the inventory level drops below the reorder point s, a reorder is triggered to fill up the inventory level to a target inventory level S. Now, the entire inventory system behavior can be described in two parameters (s, S). The task of optimally configuring these parameters has been tackled via meta-heuristic or simulation-based approaches.

Heuristic methods aim at solving optimization problems under the constraint of limited prior knowledge and limited time. Examples of heuristic methods are, e.g., genetic algorithms (Grahl et al. 2016).

Unfortunately, there are some disadvantages associated with heuristic methods: (1) they do not provide optimality guarantees, and (2) they lack general applicability. If there are changes in the supply chain network structure or in the model assumption, the method needs to be revised—whereby finding the right parameters can become a huge effort.

Simulation-based policy optimization is a frequently used approach with numerous variations (Chu et al. 2015). The main ingredients are: (1) A model simulates the inventory system network taking policy parameters as an input and outputs the corresponding performance measures. Three model categories are particularly well suited: The MEIS can be interpreted as a classic coupled tank system in control engineering turning the problem into a set of complicated differential equations which is challenging to be solved. Another option is an agent-based model as performed in (Chu et al. 2015). Each individual inventory system is modeled by the interaction of four different agents: A facility agent, an order agent, a shipment agent, and a customer agent. The characteristic of such an agent-based system can be categorized as a black-box function. The third option is a hybrid model allowing to access specific model structures while other components remain black-box functions. (2) A Monte-Carlo method estimating the expected value of the performance measure for a given set of policy parameters over multiple simulation time steps. (3) An optimization algorithm evaluating how to update the policy parameters to iteratively optimize the performance. Two drawbacks are related to this approach. Similar to the heuristic methods, the Monte-Carlo estimation performed in step (2) is computationally intense leading to poor scaling properties. Secondly, the optimization algorithm converges to a minimum, but this may be a bad local minimum.

Approximative policy optimization interprets the inventory system as a Markov Decision Process (MDP) which is covered in more detail in Sect. 4.1 (Powell 2007). MDPs are typically solved with dynamic programming (DP) approaches. However, due to the complexity of the system, these methodologies do not have the required scaling properties to be a suitable solution to a MEIO challenge. Thus, approximate dynamic programming approaches have been developed to simplify the underlying dynamic program. According to Gijsbrechts et al. (2021), these can be structured into three distinct research branches: The first branch exploits the problem structure by simplifying assumptions such as very short lead times. Another branch aggregates multiple states to a single state based on hand-crafted features. The third branch approximates the value or the policy function of the MDPs. Two famous representatives using function approximation are linear programming-approximate dynamic programming Approximate Dynamic Programming (LP-ADP) and RL with the remainder of this chapter focusing on the latter method.

4 Reinforcement Learning for Inventory Management

4.1 Markov (Decision) Processes

The operation of inventory systems is related to multiple stochastic processes (e.g., the demand and lead times) and can be modeled as a Markov Process (MP) (e.g., Broyles et al. 2010). The relevant components and quantities describing a Markov chain are states, transition probabilities, and, optionally, performance metrics describing the quality of a state. Figure 1a showcases a Markov chain modeling a highly-simplified inventory system for explanatory purposes. The state space consists of two states representing a low IOH-level (LIOH) and a high IOH-level (HIOH). Each time step, the IOH transitions from one state to another state if the IOH exceeds (e.g., through backordering) or falls below (e.g., through customer demand) a certain threshold—otherwise the system remains in the same state. The set of all transition probabilities is referred to as system dynamics and regulates the probabilities via which the state of the system changes to another state and remains constant over the full period of time to fulfill the so-called Markov property. The system dynam-

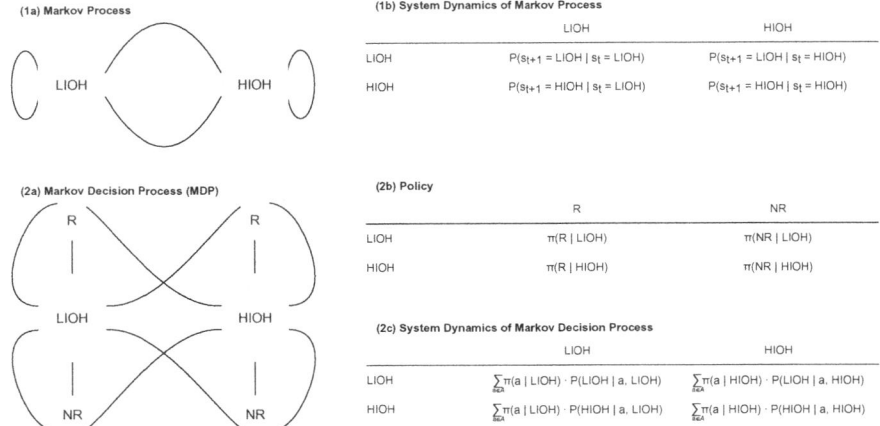

Fig. 1 (a) Describes the stochastic behavior of the inventory systems IOH in a highly-simplified way. In each time step the system can either remain in the same state or transition to the other state. Processes such as the demand may cause the IOH to decrease. Processes such as reordering may increase the IOH. The behavior of the system is described with the system dynamics displayed in (b). (d) Displays a MDP with a policy responsible for the reorder decision summarized as π (the policy) and non-controllable processes such as the demand. In each time step, an agent takes a decision (reorder or no reorder) according to the policy. The overall system behavior can thus be defined as a combination of the policy π and the system dynamics P as illustrated in (e)

ics result from processes impacting the state of the inventory systems such as the demand and reordering activities. In this use-case, we consider one performance metric describing the overall cost associated with the operation of the inventory system. To create an intuition, we can associate the state HIOH with low cost and the state LIOH with high cost. This example is arbitrary and for explanatory purposes only, but could be justified with a higher likelihood of the event of a stock-out in case of a LIOH and with shortage cost outweighing the cost for holding a high number of inventory. The expected cost over multiple time steps depends on the number of time steps and the equilibrium distribution describing how often the system is in state LIOH or HIOH respectively. The equilibrium distribution solely depends on the system dynamics and can be calculated analytically or in case of complex Markov chains estimated with a Markov Chain Monte Carlo (MCMC) simulation.

The Markov model is a suitable framework for describing stochastic systems—however, to use it as a basis for finding an optimal control strategy some extensions need to be applied: Firstly, we need to differentiate processes, which influence the system dynamics, into controllable and non-controllable processes. With regards to an inventory system, e.g., the reorder decision is a controllable process, whereby there are limited options to control the customers demand. The non-controllable processes remain referred to as system dynamics, whereby controllable processes are denoted as the policy. Modelwise, this means that each system state (LIOH/HIOH) is followed by a decision (e.g., reorder (R)/not-reorder (NR)) taken by an agent in accordance with the policy, which by itself is followed by the successor state with a dependency on the selected action and the corresponding system dynamics. With the possibility of actively intervening into the system through the policy, we have introduced the concept of an MDP (Puterman 1990). Figure 1c highlights the difference between an MP and an MDP. The policy is referred to as an optimal policy in case it minimizes the expected overall cost in a way that no other policy can be associated with a lower expected overall cost.

As mentioned, however, this is a greatly simplified model and the reality is far more complex. On the one hand, an order does not immediately lead to an increase in the inventory level. Instead, it often takes a few days for the delivery. A policy should be aware of open orders to avoid multiple reorders in a low inventory level state. In fact, this is one aspect explaining why the state space is of much higher complexity than displayed in Fig. 1: There must be a specific state for each IOH and open order combination, whereby the open-order situation can be quantified with two additional dimensions: the order timing and the order quantity. Furthermore, the inventory level should be structured in a much more fine-granular way: We need one state for each possible inventory level—instead of grouping it into low and high IOH-levels. Moreover, the action choices need to be revised: In Fig. 1 it is distinguished between reordering and not-reordering. In reality, inventory systems can select one of the multiple order quantities. These examples should provide the reader an intuition, why the real state and action space is much more complex and high-dimensional compared to the simplified variant illustrated in Fig. 1.

5 Introduction to Reinforcement Learning

RL is a promising approach to tackle inventory optimization challenges because of the following reasons:

- The policy can be represented by a deep neural network with all its related advantages: The representative capacity of neural networks is high, allowing to properly identify n-order dependencies of the input variables. In addition, deep neural networks are capable of generalization: It is not necessary to simulate every situation in a training stage (which would be computationally impossible)—it is enough to have encountered a limited set of situations and apply an action that performed well according to generalized experiences. Furthermore, the input values can be relatively unstructured and may include information that is redundant to identify the optimal control.
- RL is generally applicable to every inventory system setups—with only little expert knowledge or intense model tuning required. Intuitively speaking, the algorithm finds its own way to the optimal policy. This is the aspect that distinguishes RL from other methods that are often used in inventory optimization challenges—especially from heuristic approaches.

- Thanks to the expected reward properties which are discussed in the following section, RL is optimizing decisions considering the long-term outcomes instead of optimizing the short-term consequences. This is a very important feature of many decision-making processes.

5.1 Value-Based Methods

We already introduced five essential components in RL. The current state of a system is captured in a state vector. In an inventory system context, this state vector may include the current inventory level on hand and the open-order situation. The agent then applies a policy taking the state vector as an input and mapping it onto an action vector. This action vector may include information such as whether to reorder and how much to reorder. The action is subsequently applied to the environment causing effects regarding the systems state: The system transitions from one state to another state. Next to the state, the environment returns another signal that enables the policy to learn an optimal policy: The reward. The reward provides the agent with the information, whether the action taken in the last state was actually a good choice or not. In the case of a supply chain cost optimization use-case, this reward may represent the overall cost. However, optimizing the immediate reward can be myopic: In the short term, total cost can be reduced by not ordering and avoiding transportation cost. In the long term, this causes shortage cost due to stock-out. This example showcases the need to consider the long-term consequences of an action. A mathematical basis for integrating these long-term consequences is provided by the Bellman equation: The quality of a state is defined as the expected sum of rewards collected in the next time step and all future time steps. This value can be assigned to every state (state value v), and to every state-action pair (action value q) (Sutton and Barto 2018a).

The state value is defined as the expected reward G_t at time step t conditioned on the state S_t at time step t following a policy π.

$$V(s) = \mathbb{E}_\pi[G_t | S_t = s] \tag{1}$$

The expected reward G_t can be expressed as the sum of rewards collected at the next time step and all future time steps k. Since the uncertainty increases with increasing k, each reward can be considered with a discount factor γ^{t+k+1}, with $\gamma \in [0, 1]$. If $\gamma = 1$, each future reward is weighted equally, independent of the moment of occurrence. In contrast to this, a γ close to zero focuses on nearby rewards through masking out distant future rewards. From this it follows that

$$V(s) = \mathbb{E}_\pi \left[\sum_{k=0}^{\infty} \gamma^k R_{t+k+1} | S_t = s \right] \tag{2}$$

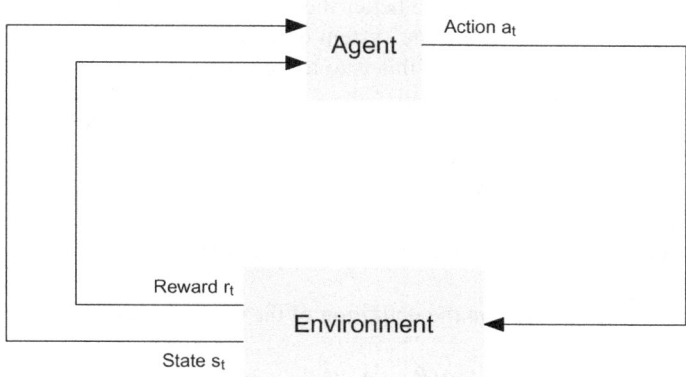

Fig. 2 An agent applies a policy $\pi(a_t|s_t)$ by mapping a state vector s_t onto an action vector a_t. The action is then applied to a system that is denoted as environment. The systems state is changed due to the agents action with the new state s_{t+1} and the reward signal r_{t+1} passed to the agent

The summand related to $k = 0$ need to be extracted from the sum, to prepare the following steps introduced in Eq. 3–5.

$$V(s) = \mathbb{E}_\pi [R_{t+1} + \gamma \sum_{k=0}^{\infty} \gamma^k R_{t+k+2} | S_t = s] \quad (3)$$

The expected reward depends on the policy $\pi(a|s)$ and the system dynamics $p(s', r|s, a)$. Taking these equations as deterministic equations helps to move them out of the expected value:

$$V(s) = \sum_a \pi(a|s) \sum_{s'} \sum_r p(s', r|s, a)[r + \gamma \mathbb{E}_\pi [\sum_{k=0}^{\infty} \gamma^k R_{t+k+2} | S_{t+1=s'}]] \quad (4)$$

A close look suffices to replace the remaining term in the expected value with a deterministic expression. The term within the expected value is similar to the expression in Eq. 2. The only difference is that Eq. 2 formulates the state value for state s, where the expected value in Eq. 4 refers to the expected value of the successor state s'. Therefore Eq. 4 can be rewritten as

$$V(s) = \sum_a \pi(a|s) \sum_{s'} \sum_r p(s', r|s, a)[r + \gamma v_\pi(s')] \quad (5)$$

Equations 1–5 and more details are summarized by Sutton and Barto (2018b). The expected reward of a state or a state-action combination depends on the policy π, system dynamics p, the reward of the current time step r, the discount factor γ, and the quality of the next state $v_\pi(s')$. While the policy-dependency seems to be intuitive (the better the decision-making, the better the expected outcome), one challenging factor must be considered: Both, the system dynamics p and the state-values may be unknown. One way to deal with this is to try to estimate the value function $V(s)$ and to further improve the estimation with every additional interaction.

Two common value function estimation methodologies exist: The Monte-Carlo (MC) method and Temporal Difference (TD) learning (Sutton and Barto 2018a). Both are based on the Bellman equation. The MC method learns in an episode-by-episode sense by accumulating the reward encountered after taking action a_t in state s_t in time step t. On the other hand, the TD method allows updating the state value in every time step.

The TD method is based on the definition of the expected reward:

$$G_t = R_{t+1} + \gamma R_{t+2} + \gamma^2 R_{t+3} + \gamma^3 R_{t+4} + \ldots \tag{6}$$

The γ for the second summand and all following summands can be factored out.

$$G_t = R_{t+1} + \gamma \cdot (R_{t+2} + \gamma^1 R_{t+3} + \gamma^2 R_{t+4} + \ldots) \tag{7}$$

As a consequence, the expected reward of the current time step t can be formulated as a sum of the reward and the expected reward of the next time step $t+1$:

$$G_t = R_{t+1} + \gamma G_{t+1} \tag{8}$$

The characteristics illustrated in Eq. 8 can be leveraged as a basis for learning: The expected rewards G_t and G_{t+1} are estimates and may originate from a model output. On the other hand, the right side of the equation incorporates a R_{t+1} which has been encountered by interacting with the environment. Thus, it can be assumed that the right side of the equation is generally more accurate and can serve as the TD target. The left-hand side can be seen as a TD prediction. The difference between TD target and TD prediction can be understood as the TD error. This underlines the analogy to supervised learning methods. A model can now be trained according to the prediction and the target, wherein RL the target is only a better estimate and in supervised learning ground truth. From another perspective, while the goal of supervised learning is to minimize the difference between the prediction and the target, RL faces an additional challenge as the target is a moving target being updated in every learning step. The procedure of updating an estimate with another estimate is called bootstrapping and explains why RL is typically more computationally intensive compared to supervised learning. Next to the learning aspect, RL is also more computationally intensive from a sampling process perspective. In supervised learning, the labeled dataset is usually static and already given at the beginning of the learning process. In the case of RL, the data is collected during the learning process through interacting with an environment.

The value function $V(S_t)$ can be updated with the temporal difference using the following update rule:

$$V(S_t) = V(S_t) + \alpha[R_{t+1} + \gamma V(S_{t+1}) - V(S_t)], \tag{9}$$

with α representing the update step length. An analogous contemplation can also be carried out for action values.

So far, it was discussed how to estimate and update a value for each specific state. Each state must be visited multiple times until the value function can estimate the corresponding value accurately. This is possible if the state of the environment can be described with a discrete vector and a limited amount of distinct states. However, if the state space becomes more numerous or even continuous, this approach becomes infeasible due to computational complexity—the computation time to develop an appropriate value estimate for each state would grow toward infinity. For this reason, we need to look at models that can approximate without much loss of performance. Is it possible to draw conclusions from an experience in one state for similar states?

This question leads inevitably to the topic of neural networks. The idea of leveraging Multi-layer Perceptrons (MLPs) as non-linear function approximators in RL was considered to be unstable until Mnih et al. (2013) proposed Deep Q-Networks (DQN). Two contributions are responsible for this breakthrough:

Firstly, the concept of an experience replay buffer was introduced: Instead of learning from experiences as they occur, they are stored in a table named replay buffer. The information stored consists of the state-action pair s_t, a_t, the corresponding reward $r_t + 1$ and the successor state $s_t + 1$. Subsequently, the experience-making and the learning process can be seen as decoupled processes as the learning takes place on the basis of randomly selected samples from the replay buffer. This procedure removes the temporal correlation between the samples and stabilizes the convergence properties.

Secondly, the concept of target networks was introduced: Two function approximators are used instead of one: (1) A target network and (2) a behavior network. The target network represents a copy of the behavior network and is used to calculate the Bellman update. The Bellman update is used to update the parameters of the behavior network. The parameters of the target network are periodically updated according to the behavior network. This concept keeps the target more stable compared to updating the target in every time step and has a stabilizing effect on the training process. Many extensions of DQN have been published such as Double Deep Q-Networks (Double DQN) (van Hasselt et al. 2015), Prioritized Experience Replay (Schaul et al. 2015), or Dueling Deep Q Networks (Dueling DQN) (Wang et al. 2016).

Despite all advantages, value-based methods cannot handle continuous action spaces without leveraging an additional optimization technique. According to (Lillicrap et al. 2015), the idea to simply discretizing a continuous action space into a fine-granular discrete action space often fails: Even small systems with little degrees of freedom are related to a sprawling action space leading to a too high sample complexity. One method to enable the control of continuous action spaces is introduced in the subsequent section.

5.2 Policy-Based Methods

This section introduces the fundamental concepts of policy-gradient algorithms. The underlying idea is to represent the policy by a parametric probability distribution $\pi_\theta(s) = P(a|s, \theta)$, where θ represents the parameters of the function approximator. In contrast to value-based methods, the output represents a probability density function that assigns a probability to each possible action. Typically, policy-gradient algorithms try to adapt the model parameter θ by estimating the gradient of the expected return G (Sutton et al. 1999). Intuitively, this can be interpreted to mean that actions that have led to a positive outcome are more likely to be selected in the future in the same or similar states:

$$\nabla_\theta \mathbb{E}_{a \sim \pi_\theta(s)}[G(a)] \qquad (10)$$

The gradient of the expected value cannot be calculated analytically due to the infinite set of state-action combinations. Alternatively, two common methods for estimating the gradient exist REINFORCE (Sutton et al. 1999) and the reparameterization trick (Kingma and Welling 2013). In the following, REINFORCE is described in three steps. Firstly, an episodes trajectory τ of length T including state, action, and reward information is collected:

$$\tau = (s_0, a_0, r_1, s_1, a_1, r_2, s_2, ..., a_T, r_{T+1}, s_{T+1}), \qquad (11)$$

with s_t, a_t, r_t representing the state, the action and the reward at time step t.

In a second step, the expected reward of each visited state is estimated: The reward r_k of each time step within the trajectory multiplied with its corresponding discount factor γ is accumulated:

$$G_t \leftarrow \sum_{k=t+1}^{T+1} \gamma^{t-k-1} R_k \qquad (12)$$

where k denotes the number of time steps ahead of t.

Finally, the model parameters are updated according to the following equation:

$$\theta_{t+1} = \theta_t + \alpha G_t \frac{\nabla_\theta \pi(A_t|S_t, \theta_t)}{\pi(A_t|S_t, \theta_t)} \qquad (13)$$

The process of estimating the expected reward is similar to the Monte-Carlo method, which is related to some advantages and disadvantages: On the one hand, the estimate is unbiased as it is based on a real trajectory. On the other hand, only a small change in the policies' parameters may change to another decision within the trajectory leading to a completely different outcome. This is why the estimation is related to a high-variance-affected credit assignment. Furthermore, the weakness of REINFORCE can be seen from another perspective: Imagine an agent taking a bad decision in time step t and good decisions in all successor time steps. REINFORCE rewards the bad action for its long-term positive outcome. This explains, why policy-based methods are considered to be less sample efficient. An alternative approach to reduce variance in the update steps can be found in the reparameterization trick. The policy $\pi(a|s, \theta)$ is reformulated by a probability distribution g_θ depending on an expected value μ_θ, standard deviation σ_θ and a stochastic value ϵ.

$$g_\theta(\epsilon) = \mu_\theta + \epsilon \sigma_\theta \tag{14}$$

This transformation decouples the expectation of the policy parameter θ and has a simplifying effect on the calculation of the gradient. Research papers demonstrate that the reparameterization trick has a variance-reducing effect (Xu et al. 2018).

5.3 Actor-Critic Methods

Actor-critic methods consist of a policy-gradient-based actor and a value-function-based critic. The actor maps the state vector onto an action vector and its corresponding update step works in the same way compared to the policy-gradient techniques introduced in Sect. 5.2. The critic takes the state vector and the action vector chosen by the actor as input and maps it on a scalar critic value. The critic value serves as a reward signal for the actor. In contrary to policy-based methods, the expected return is not estimated according to the Monte-Carlo method as denoted in Eq. 12. Instead, the expected reward is estimated according to the critic network. Two fundamental representatives are presented in Eq. 15 and in Eq. 16.

$$\nabla_\theta J(\theta) = \mathbb{E} \sum_{t=0}^{T-1} \nabla_\theta log \pi_\theta(a_t|s_t) Q(s, a) \tag{15}$$

and the Advantage Actor-Critic as denoted in Eq. 16,

$$\nabla_\theta J(\theta) = \mathbb{E} \sum_{t=0}^{T-1} \nabla_\theta log \pi_\theta(a_t|s_t) A(s, a), \tag{16}$$

whereas the advantage is defined as $A(s, a) = Q(s, a) - V(s)$. Substituting the Monte-Carlo-based expected reward estimate with a value-function-based expected reward estimate counteracts the high-variance issue related to pure policy-gradient-based methods.

The field of actor-critic methods has evolved rapidly in recent years and numerous extensions have been developed. One measure to stabilize the training process is to parallelize the learning process (Mnih et al. 2016). Asynchronous Actor-Critic (A3C) uses multiple agents with identical model architecture interacting with their own copy of the environment and collecting their own experiences. Two novel update strategies are to be considered: Firstly, the decentralized agents perform an asynchronous update step of the centralized agent using their own network gradients. These gradients contain information on how to update the network parameters based on the individual agent's experiences accumulated over multiple timesteps. Furthermore, the decentralized network parameters are substituted with the parameters of the global network. This parallelization has a stabilizing effect since the learning process is based on decoupled learning experiences similar to the experience replay buffer proposed in Sect. 5.1. In addition to the parallelization, the authors pointed out another important property: Adding the policy entropy $\mathcal{H}(\pi(\cdot|s_t))$ as a regularizer to the objective function reduces the risk of converging to a bad local optimum.

6 Evaluation

In the previous sections, many theoretical points have been discussed. Now, we are interested in how well RL works in practice. Thus, a small experiment is conducted:

The environment: A divergent 2-layer MEIS is considered. One middle warehouse orders supply from a factory and distribute supply to two leaf warehouses. The following assumptions are applied: The factory-level inventory system always has enough supplies to serve orders from a middle warehouse. The middle warehouse and the leaf warehouses can be affected by stock-out. The demand at the leaf warehouses is triggered by local wholesalers and hospitals and is modeled with a normal distribution. The demand at the middle warehouse corresponds to the orders of the two leaf warehouses (Fig. 3).

The cost. The delivery of an order is accomplished after a stochastic lead time, provided that sufficient supplies are available on the upstream level. Unserved orders due to stock-out are backlogged. The longer the waiting time for unserved demand gets, the higher is the likelihood of a buyer withdrawing the order leading to shortage cost at leaf warehouse level. Stock-out may occur at the middle warehouse level as well, however, this does not directly lead to lost sales since the middle warehouse is not directly connected to the market and therefore the shortage cost at middle warehouse level are assumed to be zero. In addition to shortage cost, there are holding cost and reordering cost. The overall cost in each time step is denoted in Eq. 17:

$$c_{total}(t) = \sum_{i \in M} c_{i,shortage}(t) + c_{i,reordering}(t) + c_{i,holding}(t), \qquad (17)$$

while each cost type is defined as

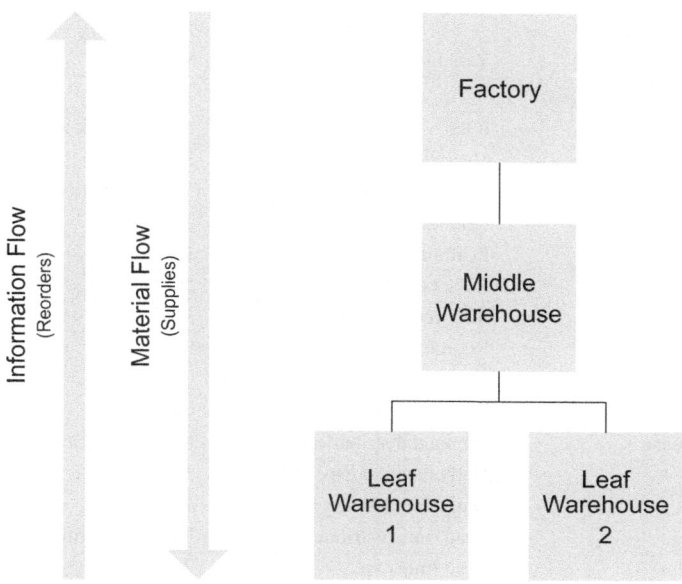

Fig. 3 The environment setup consists of four inventory systems, whereby three of them are to be controlled by the RL agent. The material flow is from top to bottom while the information flow (the orders) is from bottom to top

$$c_{shortage}(t) = k_{shortage} \cdot min(0, ioh_i(t)) \cdot ppp_i \tag{18}$$

$$c_{reordering}(t) = \begin{cases} 0, & \text{if no reorder} \\ min(c_{min.reordercost}, k_{reorder} * q_{ireorder}), & \text{otherwise} \end{cases} \tag{19}$$

$$c_{holding}(t) = k_{holding} \cdot max(0, ioh_i(t)) \cdot ppp_i, \tag{20}$$

whereby $c_{shortage}(t)$, $c_{reordering}(t)$, $c_{holding}(t)$ denote the shortage, reordering, and holding cost at time step t. $k_{shortage}$ and $k_{reorder}$ denote cost specific constants and ppp_i, $ioh_i(t)$ represent the price per product and the IOH at inventory system i and time step t. To transform the cost minimization challenge to a maximization task, we define $reward = -c_{total}$.

The state. The state vector consists of four elements per warehouse: (1) the current IOH, (2) the order quantity of the oldest open order, (3) the number of dates since the oldest open order was placed, and (4) the reorder quantity of all open orders. The respective state vectors for each warehouse are concatenated into a global state vector. The resulting state dimension is 12, if we consider three warehouses with four corresponding state dimensions.

The action. The output space is 13-dimensional as the agent has the choice to choose one out of 13 options. The first option is that no warehouse orders—all remaining

Table 1 Specifications

Category	Variable	Value
Environment specifications		
Factory	IOH	Always sufficient
	Overall cost	Always zero
Middle warehouse	Lead time distribution	Normal distributed
	Lead time exp. [days]	2
	Lead time std. [days]	1
	Price per product [CHF]	50
	Min. reorder cost [CHF]	1000
	Reorder cost constant [CHF]	0
	Shortage cost constant [CHF]	0
	Holding cost constant [CHF]	0.1
Leaf warehouse	Demand distribution	Normal distributed
	Daily demand exp. [days]	3300
	Daily demand var. [days]	100
	Lead time distribution	Normal distributed
	Lead time exp.	2
	Lead time std.	1
	Price per product [CHF]	100
	Min. reorder cost [CHF]	5000
	Reorder cost constant [CHF]	0.5
	Shortage cost constant [CHF]	10
	Holding cost constant [CHF]	0.1
	Max. backlog duration [days]	7
Agent Specifications		
Agent	Approach	A3C
Model	Actor model	FCMLP
	Actor model: No. layers	3
	Actor model: No. neurons per layer	64
	Critic model	FCMLP
	Critic model: No. layers	3
	Critic model: No. neurons per layer	64
Training	No. episodes	500K
	No. time steps per training episode	365
	Optimizer	Adam
	Learning rate	0.0001
	Discount factor (γ)	0.99

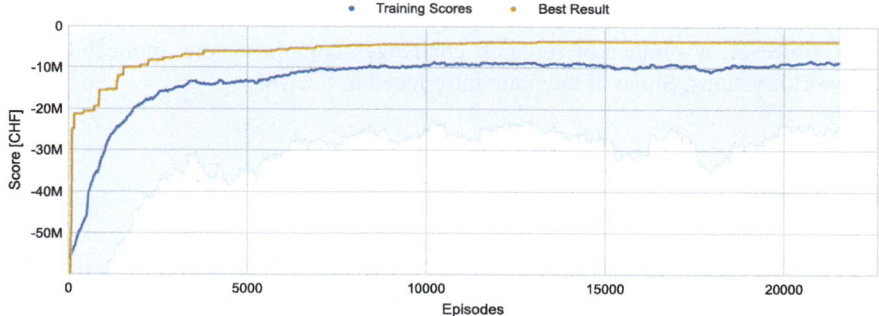

Fig. 4 Visualization of the RL model training performance. The blue line reflects the smoothed scores, i.e., smoothed means of negative cost achieved during training. The blue shaded area denotes standard deviations of these score values. The orange line denotes the overall best result obtained so far during training

options represent the situation that only one warehouse can reorder at the same time step while the reorder quantity can be small, medium, large or extra large.

The agent. The optimal policy is developed with the A3C approach. The model consists of two Fully-Connected Multi-layer Perceptrons (FCMLPs)—one for the actor and one for the critic. Further model and training hyperparameters are listed in Table 1.

Figure 4 illustrates the training performance progress. It can be observed that the cost converges to an annual cost of below 10M CHF. However, the cost fluctuation remains on a high level. This issue is further discussed in Sect. 8.

7 Discussion of Results

Section 6 describes an experiment on MEIO with the A3C approach. The RL agent is capable of learning a reorder policy with minimized overall cost for a small, divergent multi-echelon network. It remains a research question to be answered, how good the performance is compared to other optimization methods. Gijsbrechts et al. (2021) performed a similar experiment by comparing two different kinds of base-stock policies: One base-stock policy is associated with constant base-stock values, while the other is state-dependent, whereby the corresponding base-stock values are selected by an A3C agent. The experiment shows, that the A3C-based approach outperformed the other approach by 9–12% less overall cost. On the other hand, the experiment performed in Sect. 6 shows, that the training converges to a minimized cost—on the other hand, the variance of the performance remains comparably high and no performance guarantees are given. In summary, RL shows promising results for inventory management tasks and many other sequential decision-making use-cases, however, a number of research challenges complicate the applicability to real-world systems. Some of them are introduced in Sect. 8.

8 Outlook

The most important open research challenge in the field of MEIO with RL is to make DRL-agents reliable and trustworthy. The experiment presented in Sect. 6 demonstrated, that A3C learns an optimized policy. Numerous very good runs alternate with a few very bad runs. With a deep neural network as a function approximator, the policy remains a black-box function with limited interpretability and thus no performance guarantees can be given. It still needs to be clarified how the trustworthiness of DRL can be increased and guaranteed. One research branch targeting this is analysed by Garcýa and Fernández (2015). Another aspect targets the environment: This is based on simplifying assumptions that often do not match the properties of real supply chains. One example is perishability and associated write-off cost in case of product expiry. Other examples are physical constraints (e.g., constraint workloads regarding the number of processible orders) or legal constraints (e.g., fixed safety stock regulations). Furthermore, the demand, which is sampled from a normal distribution in Sect. 6 may oversimplify the real demand characteristics and make the simulation-based learned policy not suitable for the application in real-world settings. Special events (e.g., a pandemic leading to demand artifacts) or low demands in the rare disease area facing high uncertainty may lead to poor results in reality if they are not considered in the simulation. In future research efforts, this characteristic could be captured with temporal point processes (e.g., Reinhart 2018).

9 Conclusions

DRL is a rapidly evolving research field. Experiments show two-fold results: On the one hand, DRL learns an optimized reorder policy with a low overall cost. On the other hand, the performance variance is relatively high with many good episodes alternating with some poor episode results. This makes DRL a promising approach to optimizing inventory management in the future—however, with the current lack of performance stability, DRL inventory management requirements and state of the art are too remote to be considered as a serious alternative for application to real-world supply chains.

10 Acronyms

AI	Artificial Intelligence
A3C	Asynchronous Actor-Critic
FCMLP	Fully-Connected Multi-layer Perceptron
DDPG	Deep Deterministic Policy Gradient
DPG	Deterministic Policy Gradient
DQN	Deep Q-Networks
DRL	Deep Reinforcement Learning
LP-ADP	Approximate Dynamic Programming
MARL	Multi-agent Reinforcement Learning
MC	Monte Carlo
MDP	Markov Decision Process
MDP's	Markov Decision Processes
MEIO	Multi-echelon Inventory Optimization
MEIS	Multi-echelon Inventory Systems
MLP	Multi-layer Perceptron
MLP's	Multi-layer Perceptrons
PPO	Proximity Optimisation
RL	Reinforcement Learning
SAC	Soft Actor-Critic
TD	Temporal Difference
TRPO	Trust Region Policy Optimisation
Double DQN	Double Deep Q Networks
Dueling DQN	Dueling Deep Q Networks
IOH	Inventory on hand
T	Reorder timing
Q	Reorder quantity
MCMC	Markov Chain Monte Carlo
LIOH	Low IOH-level
HIOH	high IOH-level
R	Reorder
NR	Not-reorder
MP	Markov Process

References

Broyles, J. R., Cochran, J. K., & Montgomery, D. C. (2010). A statistical markov chain approximation of transient hospital inpatient inventory. *European Journal of Operational Research, 207*(3), 1645–1657.

Chu, Y., You, F., Wassick, J. M., & Agarwal, A. (2015). Simulation-based optimization framework for multi-echelon inventory systems under uncertainty. *Computers & Chemical Engineering, 73*, 1–16.

Clark, A. J., & Scarf, H. (1960). Optimal policies for a multi-echelon inventory problem. *Management Science, 6*(4), 475–490.

De Kok, T., Grob, C., Laumanns, M., Minner, S., Rambau, J., & Schade, K. (2018). A typology and literature review on stochastic multi-echelon inventory models. *European Journal of Operational Research, 269*(3), 955–983.

Garcýa, J., & Fernández, F. (2015). A comprehensive survey on safe reinforcement learning. *Journal of Machine Learning Research, 16*(1), 1437–1480.

Gijsbrechts, J., Boute, R. N., Van Mieghem, J. A., & Zhang, D. (2021). Can deep reinforcement learning improve inventory management? performance on dual sourcing, lost sales and multi-echelon problems. *Manufacturing & Service Operations Management*.

Grahl, J., Minner, S., & Dittmar, D. (2016). Meta-heuristics for placing strategic safety stock in multi-echelon inventory with differentiated service times. *Annals of Operations Research, 242*(2), 489–504.

Kingma, D. P., & Welling, M. (2013). Auto-encoding variational bayes. arXiv:1312.6114.

Kiran, B. R., Sobh, I., Talpaert, V., Mannion, P., Sallab, A. A. A., Yogamani, S., & Pérez, P. (2020). *Deep reinforcement learning for autonomous driving: A survey.* arXiv:2002.00444

Lillicrap, T. P., Hunt, J. J., Pritzel, A., Heess, N., Erez, T., Tassa, Y., . . . Wierstra, D. (2015). Continuous control with deep reinforcement learning. arXiv:1509.02971.

Mnih, V., Badia, A. P., Mirza, M., Graves, A., Lillicrap, T., Harley, T., . . . Kavukcuoglu, K. (2016). Asynchronous methods for deep reinforcement learning. In *International conference on machine learning* (pp. 1928–1937).

Mnih, V., Kavukcuoglu, K., Silver, D., Graves, A., Antonoglou, I., Wierstra, D., & Riedmiller, M. A. (2013). Playing atari with deep reinforcement learning. Retrieved from arXiv:1312.5602

Powell, W. B. (2007). *Approximate dynamic programming: Solving the curses of dimensionality* (Vol. 703). Wiley.

Puterman, M. L. (1990). Markov decision processes. *Handbooks in Operations Research and Management Science, 2*, 331–434.

Reinhart, A. (2018). A review of self-exciting spatio-temporal point processes and their applications. *Statistical Science, 33*(3), 299–318.

Ribba, B., Dudal, S., Lavé, T., & Peck, R. W. (2020). Model-informed artificial intelligence: Reinforcement learning for precision dosing. *Clinical Pharmacology & Therapeutics, 107*(4), 853–857.

Schaul, T., Quan, J., Antonoglou, I., & Silver, D. (2015). Prioritized experience replay. arXiv:1511.05952.

Silver, D., Huang, A., Maddison, C. J., Guez, A., Sifre, L., Van Den Driessche, G., . . . others (2016). Mastering the game of go with deep neural networks and tree search. *Nature, 529*(7587), 484–489.

Silver, D., Hubert, T., Schrittwieser, J., Antonoglou, I., Lai, M., Guez, A., . . . Hassabis, D. (2017). *Mastering chess and shogi by self-play with a general reinforcement learning algorithm.* arxiv:1712.01815, https://doi.org/10.48550/ARXIV.1712.01815

Sutton, R. S., & Barto, A. G. (2018a). *Reinforcement learning: An introduction* (2nd Ed.). The MIT Press. Retrieved from http://incompleteideas.net/book/the-book-2nd.html

Sutton, R. S., & Barto, A. G. (2018b). *Reinforcement learning: An introduction.* MIT press.

Sutton, R. S., McAllester, D. A., Singh, S. P., Mansour, Y., et al. (1999). Policy gradient methods for reinforcement learning with function approximation. In *Nips* (Vol. 99, pp. 1057–1063).

van Hasselt, H., Guez, A., & Silver, D. (2015). Deep reinforcement learning with double q-learning. arxiv:1509.06461.

Wang, Z., Schaul, T., Hessel, M., Hasselt, H., Lanctot, M., & Freitas, N. (2016). Dueling network architectures for deep reinforcement learning. In *International conference on machine learning* (pp. 1995–2003).

Xu, M., Quiroz, M., Kohn, R., & Sisson, S. A. (2018). *Variance reduction properties of the reparameterization trick*.

Patric Hammler is a Ph.D. student at University of Berne and a Data Scientist at the Roche Pharma International Data and Analytics chapter. His research focuses on Deep Reinforcement Learning and its applications in the areas of Supply Chain Optimization and Building Automation.

Nicolas Riesterer Ph.D. student in Computer Science with a strong focus on AI and Machine Learning. Currently working at Roche as a Data Scientist in the Data & Analytics Department.

Gang Mu Holds a Ph.D. degree in mathematics. Comprehensive experiences to connect Mathematics, Healthcare and Technology together driving impacts and outcomes for patients and healthcare systems. Founded Swiss Network for Mathematics in Industry. Head of AI for Partnerships at Roche and Visiting Research Scholar at the University of Zurich.

Torsten Braun Professor in Computer Science at University of Bern and director of the Research Group "Communication and Distributed Systems". Director of the Institute of Computer Science, former Vice Dean of the Faculty of Science, former Vice President of SWITCH foundation, partly as interim President.

Open Access This chapter is licensed under the terms of the Creative Commons Attribution 4.0 International License (http://creativecommons.org/licenses/by/4.0/), which permits use, sharing, adaptation, distribution and reproduction in any medium or format, as long as you give appropriate credit to the original author(s) and the source, provide a link to the Creative Commons license and indicate if changes were made.

The images or other third party material in this chapter are included in the chapter's Creative Commons license, unless indicated otherwise in a credit line to the material. If material is not included in the chapter's Creative Commons license and your intended use is not permitted by statutory regulation or exceeds the permitted use, you will need to obtain permission directly from the copyright holder.

Specialized Quantitative Tools in the Life Science Industry

An Invitation to Stochastic Differential Equations in Healthcare

Dimitri Breda, Jung Kyu Canci, and Raffaele D'Ambrosio

1 Introduction

A typical family of differential/integral equations studied in healtcare or in finance is the following one:

$$V_t = V_0 + \int_0^t \alpha(s, V_s)ds + \int_0^t \sigma(s, V_s)dW_s, \tag{1}$$

where $\mathbf{V} = (V_t : 0 \leq t \leq T)$ is a d-dimensional quantity, which for example could represent the values of assets in a portfolio and W is a Brownian motion (briefly introduced in the next section).

The above type of equations are an important tool in mathematical finance.

Equation in (1) is the integral version of the equation

$$\frac{dV_t}{dt} = \alpha(t, V_t) + \sigma(t, V_t)\frac{dW}{dt}. \tag{2}$$

Indeed by operating the integral operator $f(t) \mapsto \int_0^t f(s)ds$ on the left of (2), we obtain

---

D. Breda (✉)
University of Udine, Udine, Italy
e-mail: dimitri.breda@uniud.it

J. K. Canci
Lucerne University of Applied Sciences and Arts, Lucerne, Switzerland
e-mail: jungkyu.canci@hslu.ch

R. D'Ambrosio
University of L'Aquila, L'Aquila, Italy
e-mail: raffaele.dambrosio@univaq.it

$$\int_0^t \frac{dV_s}{ds} ds = \int_0^t dV_s = V_t - V_0.$$

By operating the integral operator to the right-hand side of (2), we obtain

$$\int_0^t \left(\alpha(s, V_s) + \sigma(s, V_s) \frac{dW}{ds} \right) ds = \int_0^t b(s, V_s) ds + \int_0^t \sigma(s, V_s) dW.$$

Therefore, we obtain

$$V_t - V_0 = \int_0^t \alpha(s, VS_s) ds + \int_0^t \sigma(s, VS_s) dW,$$

that is Eq. (1).

Now we multiply both hand sides of (2) by dt to obtain the differential equation

$$dV_t = \alpha(t, V_t) dt + \sigma(t, V_t) dW.$$

1.1 Brownian Motions

There are several ways to define what is a Brownian motion. We present the one contained in (Gobet 2022, Definition 4.1.1 at page 120)

Definition 1.1 (*Brownian Motion in Dimension 1*) A Brownian motion in dimension 1 is a continuous-time stochastic process $\{W_t; t \geq 0\}$ with a continuous path, such that

- $W_0 = 0$;
- the time increment $W_t - W_s$ ($0 \leq s < t$) has the Gaussian distribution with zero mean and variance $(t - s)$;
- for any $0 = t_0 < t_1 < \ldots < t_n$, the increments $\{W_{t_{i+1}} - W_{t_i} \mid 0 \leq i \leq n - 1\}$ are independent.

A discretization of a Brownian motion is a random walk, or in other words a Brownian motion is the continuous version of a random walk.

One denotes a Brownian motion with the letter W because the mathematical theory of Brownian motions was formalized and studied by Wiener in the middle of the twentieth century. The name "Brownian" comes from the botanist Robert Brown, who used this model of motion (without formalizing it) for describing the movement of a particle (pollen) in water.

Nowadays, Brownian motion is used in finance (e.g., for evaluating assets, portfolio, gains, wealth...) and healtcare, see for example, Donnet and Samson (2013) or Ferrante et al. (2005) in the case of pharmacokinetic/pharmacodynamic models (aka PK/PC models).

Brownian motion has the advantage to be a good tool for modeling in finance by using mathematical models, and so via equations.

The mathematical disadvantage is that a Brownian motion, considered as a function of the time t is continuous but not always derivable. But, this disadvantage turns into an advantage, because it may cover a large set of examples in real-world problems.

Thus, the integral $\int_a^b f(t) \mathrm{d}W$ in the case of a Brownian motion W has no meaning in the traditional sense as Riemann–Stieltjes Integral. The notion of Ito's integral gives a definition for the integral $\int_a^b f(t) \mathrm{d}W$ in the case of a Brownian motion W.

1.2 Ito's Integral and Solutions of Geometric Brownian Motions (GBM)

In this section, we show the definition of Ito's integral and some of its applications. Everything is considered in dimension 1; thus, every function considered is a function of the time t and assumes values in $\mathbb{R}$ (real numbers). The extension of the case where the outputs of our functions are d-dimensional vectors in $\mathbb{R}^d$ is straightforward.

Definition 1.2 (*Ito's Integral*) Let f be a continuous function with respect to time t on an interval $[a, b]$. Assume that W is a Brownian motion. Then we define the Ito's integral of f with respect to W as

$$\int_a^b f(t) \mathrm{d}W = \lim_{n \to \infty} \sum_{i=0}^{n-1} f(t_i)(W_{t_{i+1}} - W_{t_i}),$$

where $t_0 = a < t_1 < \ldots < t_{n-1} < t_n = b$ represent the endpoints of a subdivision of the interval $[a, b]$ in n subintervals.

One can see that the limit converges in probability.

The condition f continuous can be weakened. Since we are considering t belonging to intervals, we are considering the σ-algebra of borelian on $\mathbb{R}$ (i.e., the Borel algebra, which is generated by open sets in $\mathbb{R}$). In Definition 1.2, it is enough to ask that f is Borel-measurable (preimages of Borel sets are Borel sets). Continuous functions are Borel-measurable, but there are Borel-measurable functions that are not continuous. For example, piecewise functions are Borel-measurable. A more "exotic" example is the indicator function $\chi_\mathbb{Q}$ (which is 1 in the rational numbers $\mathbb{Q}$ and zero otherwise), it is a Borel-measurable function even if it is highly non-continuous.

One can find the definition of Ito's integral in (Shreve 2004, Sect. 4.3, precisely on page 134). In Shreve (2004) the assumption on f is that the function f is an adapted stochastic process, that can be essentially translated into being a Borel-measurable function over time. Alternatively, one can read (Gobet 2022, Sect. 4.2, pages 132–135).

Roughly speaking, Ito's integral is defined as a Riemann Integral, where we substitute a "linear deterministic variable x" with a stochastic one. So, in other words, we can say that an Ito's integral is a limit of a sequence of *stochastic Riemann's sums* (or in case a *stochastic Legesque Integral*). But note that, in the definition of Ito's integral, one always take a "left" stochastic Riemann's sum.

In Eq. (1), the integral $\int_0^t \alpha(s, V_s) ds$ is a deterministic one (i.e., no random variable appears), thus this is Riemann's integral (or Lebesgue's one). The integral $\int_0^t \sigma(s, V_s) dW_s$ is an Ito's one.

For a given realization (or simulation) of the Brownian motion W_t, it is possible to determine an approximation for V_t. But sometimes, an exact value for the deterministic integral or an exact value of the Ito's integral are not determinable. It is always possible to give an approximated value for the integrals.

For some special cases, it is possible to find exact solutions of the equation in (1), for example, in the case of a Brownian motion, where α and σ are constant. If so, the function V_t is (1) is called *geometric Brownian motion*.

As a straightforward application of Ito's formula (see for example, (Shreve 2004, Theorem 4.4.1, p. 138), or (Gobet 2022, Theorem 4.2.5 p. 137) for a more general formulation) proves that

$$V_t = V_0 \cdot e^{\left(\alpha - \frac{\sigma^2}{2}\right)t + \sigma \cdot W_t}$$

is the solution of the Eq. (1) in the case V_t is a geometric Brownian motion.

1.3 Existence of Solutions of Stochastic Differential Equations

Under certain hypotheses, Eq. (1) admits a solution, which is unique. This was proven by Pardoux and Peng in Pardoux and Peng (1990).

Theorem 1.1 (Pardoux and Peng 1990) *Let W be a Brownian motion and α, σ the functions of Eq. (1). Let $T > 0$ be a given real number. Suppose that α, σ are continuous functions and there exist a constant $C_{\alpha,\sigma}$ (depending of α and σ) such that, for all $t \in [0, T]$ and x, y, we have*

- $|\alpha(t, x) - \alpha(t, y)| + |\sigma(t, x) - \sigma(t, y)| \leq C_{\alpha,\sigma} |x - y|;$
- $\sup_{0 \leq t \leq T} (|\alpha(t, 0)| + |\sigma(t, 0)|) \leq C_{\alpha,\sigma}.$

Then, for each $V_0 \in \mathbb{R}$, there exists a unique solution of Eq. (1).

Unfortunately, the above theorem does not give a method for determining the solution for the Eq. (1). In some cases, for example, for geometric Brownian motion, the solution is explicitly determinable. But in general, there is no general approach for solving all equations of the shape as in (1). Only in a few cases, we are able to apply an algorithm or formula for solving exactly a stochastic differential equation.

The conditions contained in the above theorem are "uniform Lipschitz conditions". This is not so surprising. For deterministic equations and so ordinary differential equations, the Picard–Lindelöff Theorem requires Lipschitz condition as well (recall that the Picard–Lindelöff Theorem gives sufficient conditions for the existence and uniqueness of ordinary first-degree differential equations). Actually, the proof in the stochastic case of SDEs looks like the analogous of the ODEs case, where there is a somewhat fixed point theorem. In the Picard–Lindelöff Theorem, the Banach–Caccipoli's fixed point theorem is used.

In Theorem 1.1, they use a fixed point theorem. The proof could inspire a way to find a method for finding a numerical approximation of the solution, which is not so efficient. For more details about the proof of Theorem 1.1, see for example, Pardoux and Peng (1990) or Ma and Zhang (2002).

2 Numerical Methods for SDEs

Theorem 1.1 only provides the assumptions that Eq. (1), equipped by the initial value $V(0) = V_0$, for the existence and uniqueness of its solution. However, this result is only qualitative and does not provide any methodological tool to compute such a solution. It is also worth highlighting that analytical solutions to SDEs can only be provided for a limited amount of simple cases; the most realistic ones, due to their complex structure, can only be numerically solved. The design and the analysis of reliable, efficient, and accurate numerical methods for SDEs have attracted the literature of the last couple of decades. A very brief—and far from being exhaustive—list of references contains Bouchard and Touzi (2004), Gobet et al. (2005), Arnold (1974), Buckwar and D'Ambrosio (2021), Buckwar et al. (2005), Buckwar et al. (2010), Burrage and Burrage (2012), Burrage and Burrage (2014), Burrage and Tian (2004), Chen et al. (2020), D'Ambrosio and Giovacchino (2021a), D'Ambrosio and Giovacchino (2021b), D'Ambrosio and Scalone (2021b), Fang and Giles (2020), Vom Scheidt (1989), Gardiner (2004), Higham (2001), Higham (2000), Higham (2021), Higham and Kloeden (2005), Hutzenthaler and Jentzen (2015), Hutzenthaler et al. (2011), Kloeden (2002), Kloeden and Platen (1992), Ma et al. (2012), Mao (2007), Melbø and Higham (2004), Saito and Mitsui (1996), Milstein (1994), Milstein et al. (2002), Misawa (2000), Neuenkirch et al. (2019), Rössler (2010), Rössler (2009), Rössler (2006), Ruemelin (1982), Abdulle et al. (2014), Chartier et al. (2014), Abdulle et al. (2013), Abdulle et al. (2012), Cohen and Vilmart (2022), Chen et al. (2016), Cohen and Dujardin (2014), Cohen (2012), de la Cruz (2020), de la Cruz et al. (2019), de la Cruz et al. (2017), Jimenez and de la Cruz Cancino (2012), de la Cruz et al. (2010) and references therein. In the remainder of the treatise, we aim to provide a few examples of such methods taken from the most famous ones. Anyway, before listing specific methods, let us recover two fundamental notions in stochastic numerics, that provide a measure for the accuracy of the corresponding scheme: the concepts of strong and weak convergence.

Given the uniform partition

$$\mathscr{I}_h = \{t_n = nh, \quad n = 0, 1, \ldots, N = T/h\} \tag{3}$$

of the interval $I = [0, T]$; let us denote by V_n the numerical solution to (1) in the point t_n, computed by a given numerical scheme. The main question is the following: how far is the numerical solution from the exact one? Does the gap between them collapse as N goes to infinity? The following definition (see Kloeden and Platen 1992; Higham 2001, 2021 and references therein) helps clarify this scenario.

Definition 2.1 Given a numerical method computing $X_n \approx X(t_n)$, with $t_n \in \mathscr{I}_h$, we say that the method

- is strongly convergent with strong order p if there exist three positive constants C, p and h^* such that

$$\sup_{t_n \in \mathscr{I}_h} \mathbb{E}\left[\left|X_n - X(t_n)\right|\right] \leq Ch^p, \tag{4}$$

for any $h \leq h^*$. The strong order p is the biggest number such that (4) holds true;
- chosen a functional space S and given $\Phi \in S$, we say that the methods are weakly convergent with weak order q if there exist three positive constants D, q, and $\widetilde{h}$ such that

$$\sup_{t_n \in \mathscr{I}_h} \left|\mathbb{E}[\Phi(X_n)] - \mathbb{E}[\Phi(X(t_n))]\right| \leq Dh^q, \tag{5}$$

for any $h \leq \widetilde{h}$. The weak order q is the biggest number such that (5) holds true.

Usually, S is the space of algebraic polynomials of degree q. In other terms, Definition 2.1 gives two possible measures for the accuracy of a stochastic numerical method: the expected error (strong convergence) and the gap between the expectations of the numerical and the exact solutions (weak convergence). One can prove that strong convergence implies weak convergence, while the vice versa is generally not true.

2.1 Euler–Maruyama Method

The simplest numerical method for deterministic differential equations $y'(t) = f(t, y(t))$ (i.e., the famous *Euler* method) is obtained by means of truncated Taylor series arguments as follows. First of all, let us compute

$$y(t_{n+1}) = y(t_n + h) = y(t_n) + hy'(t_n) + \mathscr{O}(h^2)$$
$$= y(t_n) + hf(t_n, y(t_n)) + \mathscr{O}(h^2).$$

Neglecting the term $\mathscr{O}(h^2)$ and reading the corresponding approximate equality among exact values as an exact equality among approximate values yields

$$y_{n+1} = y_n + hf(t_n, y_n), \quad n = 0, 1, \ldots, N. \tag{6}$$

Last equality provides the nonlinear difference equation defining the Euler method. Clearly, solving such a nonlinear difference equation is not simpler than solving the original ODE and, indeed, it is used to start a step-by-step procedure for the pointwise computation of the numerical solution.

This approach does not directly apply to SDEs, due to the nowhere differentiability of the involved stochastic processes. However, Taylor expansions generalize in the so-called *Ito–Taylor expansions*, thanks to the Ito formula. Indeed, specifying (1) to a subinterval $[t_n, t_{n+1})$ of the discretization $\mathscr{I}_h$ leads to

$$V(t_{n+1}) = V(t_n) + \int_{t_n}^{t_{n+1}} \alpha(s, V(s)) ds + \int_{t_n}^{t_{n+1}} \sigma(s, V(s)) dW(s).$$

Computing $\alpha(s, V(s))$ and $\sigma(s, V(s))$ by the Ito formula and considering only the very first term, i.e., $\alpha(s, V(s)) \approx \alpha(t_n, V(t_n))$ and $\beta(s, V(s)) \approx \alpha(t_{n+1}, V(t_{n+1}))$, yields

$$V(t_{n+1}) \approx V(t_n) + h\alpha(t_n, V(t_n)) + \sigma(t_n, V(t_n))\Delta W_n,$$

with $\Delta W_n = W(t_{n+1}) - W(t_n)$ (it is worth recalling that, by definition of the Wiener process, ΔW_n is a normal random variable with 0 mean and variance h). Recasting this approximate equality among exact values as an exact equality among approximate values get

$$V_{n+1} = V_n + \alpha(t_n, V_n)h + \sigma(t_n, V_n)\Delta W_n. \tag{7}$$

Equation (7) gives the so-called *Euler–Maruyama* method for SDEs. Clearly, if the diffusion coefficient σ is identically zero (i.e., the problem is deterministic), then the Euler–Maruyama method (7) recovers the deterministic Euler method (6).

One can prove that the strong order of the Euler–Maruyama method is $p = 1/2$, while its weak order is $q = 1$. For a formal proof of the strong and weak convergence of the Euler–Maruyama method, the interested reader can refer to Higham (2021) and references therein.

2.2 ϑ–Maruyama Methods

The Euler method can be merged into a larger family of methods, well-known as ϑ–*Maruyama methods* (see, for instance, Buckwar and Sickenberger 2011; D'Ambrosio and Giovacchino 2021a; D'Ambrosio and Scalone 2021a; Higham 2000, 2021 and references therein). The starting point to develop ϑ-Maruyama methods is similar to that for Euler–Maruyama method, i.e.,

$$V(t_{n+1}) = V(t_n) + \int_{t_n}^{t_{n+1}} \alpha(s, V(s)) ds + \int_{t_n}^{t_{n+1}} \sigma(s, V(s)) dW(s),$$

but the deterministic integral is approximated by the following quadrature formula:

$$\int_{t_n}^{t_{n+1}} f(t, X(t)) dt = \left[(1 - \vartheta) f(t_n, X(t_n)) + \vartheta f(t_{n+1}, X(t_{n+1})) \right] \Delta t.$$

Then, the corresponding approximate solution to (1) is given by

$$V_{n+1} = V_n + (1 - \vartheta) h \alpha(t_n, V_n) + \vartheta h \alpha(t_{n+1}, V_{n+1}) + \sigma(t_n, X_n) \Delta W_n, \quad (8)$$

$n = 0, 1, \ldots, N - 1$. Equation (8) collects the family of ϑ-Maruyama methods. Relevant cases are given for $\vartheta = 0$, leading to Euler–Maruyama method, for $\vartheta = 1/2$, leading to the *stochastic trapezoidal method* and $\vartheta = 1$, leading to the *stochastic implicit Euler method*. It has been proved (see Higham 2021 and reference therein) that all ϑ-Maruyama methods have a strong order $1/2$ and weak order 1, as it happens for Euler–Maruyama method. However, even if the accuracy is the same, selecting proper values of ϑ may provide very good stability improvements (Buckwar and Sickenberger 2011; Higham 2000; D'Ambrosio and Giovacchino 2021a).

2.3 Stochastic Runge–Kutta Methods

The relevant class of Runge–Kutta methods has its own stochastic counterpart in the family of *stochastic Runge–Kutta methods* (SRK; see, for instance, Buckwar et al. 2010; Burrage and Burrage 2012, 2014; Burrage and Tian 2004; D'Ambrosio and Giovacchino 2021b; Ma et al. 2012; Rössler 2010, 2009, 2006 and references therein). In this section, we look at SRK methods for (1) as the stochastic perturbation of deterministic Runge–Kutta methods as follows:

$$V_{n+1} = V_n + h \sum_{i=1}^{s} b_i \alpha(t_n + c_i h, \widehat{V}_i) + \Delta W_n \sum_{i=1}^{s} d_i \sigma(t_i + c_i \Delta t, \widehat{V}_i), \quad (9)$$

with

$$\widehat{V}_i = V_n + h \sum_{j=1}^{s} a_{ij} \alpha(t_n + c_j h, \widehat{V}_j) + \Delta W_n \sum_{j=1}^{s} \gamma_{ij} \sigma(t_i + c_j \Delta t, \widehat{V}_j), \quad i = 1, 2, \ldots, s. \quad (10)$$

The number s appearing above is the number of internal stages and an s-stage SRK method in the form (9)–(10) is uniquely identified by its coefficients b_i, d_i, a_{ij}, and γ_{ij}, $i, j = 1, 2, \ldots, s$, that can be collected in the following *Butcher tableau*:

$$\begin{array}{c|cccc|cccc}
c_1 & a_{11} & a_{12} & \ldots & a_{1s} & \gamma_{11} & \gamma_{12} & \ldots & \gamma_{1s} \\
c_2 & a_{21} & a_{22} & \ldots & a_{2s} & \gamma_{21} & \gamma_{22} & \ldots & \gamma_{2s} \\
\vdots & \vdots & \vdots & \vdots & \vdots & \vdots & \vdots & \vdots & \vdots \\
c_s & a_{s1} & a_{s2} & \ldots & a_{ss} & \gamma_{s1} & \gamma_{s2} & \ldots & \gamma_{ss} \\
\hline
 & b_1 & b_2 & \ldots & b_s & d_1 & d_2 & \ldots & d_s
\end{array} \qquad (11)$$

The internal stages $\widehat{V}_i$, $i = 1, 2, \ldots, s$, provide approximations to $V(t_n + c_i h)$ and the way they relate to each other makes the corresponding methods implicit or explicit. Explicit methods, i.e., with $a_{ij} = \gamma_{ij} = 0$ for $j \geq i$, have been developed in Vom Scheidt (1989); Ruemelin (1982) and provided the condition for the mean-square convergence

$$\sum_{i=1}^{s} b_i = \sum_{i=1}^{s} d_i = 1.$$

Further results, including the development and analysis of implicit methods, have been investigated in Buckwar et al. (2010), Burrage and Burrage (2012), Burrage and Tian (2004), D'Ambrosio and Giovacchino (2021b), Rössler (2010), Rössler (2009), Rössler (2006) and references therein. A two-step generalization of stochastic Runge–Kutta methods has been introduced and analyzed in D'Ambrosio and Scalone (2021b).

3 A Numerical Evidence on PK/PD Models

Let us now provide a brief selection of numerical experiments showing the effectiveness of the aforementioned approaches. The test is focused on the application of ϑ methods (8) to the following pharmacokinetic/pharmacodynamic (PK/PD) models, given by the stochastic Gompertz PD model of the bacterial count under the effect of an antibiotic Ferrante et al. (2005)

$$dN_t = (r - b\log(N_t) - kC_t)N_t dt + \gamma N_t dW_t, \qquad (12)$$

where r is the intrinsic growth rate, b is the growth deceleration rate, k is the bacterial effect of the drug, and γ is a constant parameter. This equation is coupled with a deterministic constraint on the antibiotic concentration C_t, given by

$$C_t = \frac{Dk_a}{V(k_a - k_e)} \left(e^{-k_e t} - e^{-k_a t} \right), \qquad (13)$$

where D is the dose of antibiotic, V the volume of distribution, k_a and k_e are the absorption and elimination constants, respectively. The profile of the numerical solu-

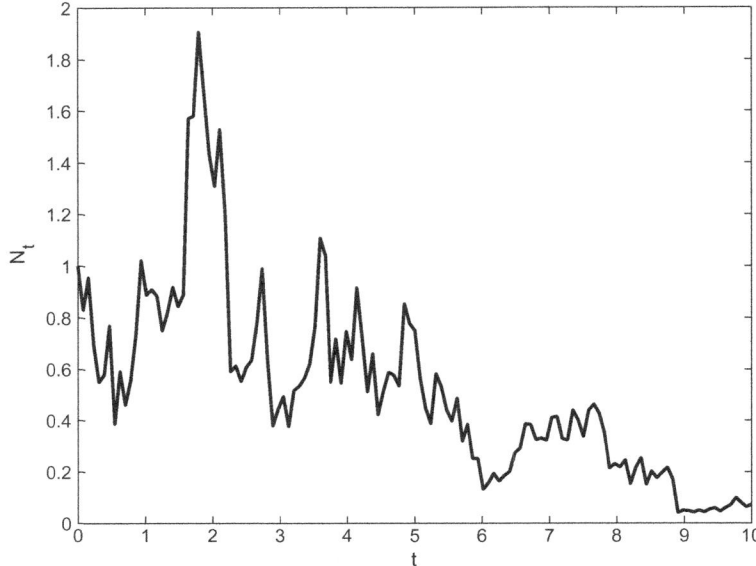

Fig. 1 Numerical solution of the PK/PD model (12)–(13), computed by the ϑ method (8) with $\vartheta = 1/2$, with $D = 1, k_a = 0.1, k_e = 0.2, V = 10, r = 0.1, b = 0.2, k = 0.1, \gamma = 1$

tion of the overall PK/PD model (12)–(13) is depicted in Fig. 1. Such a graph is obtained by applying the ϑ method (8) with $\vartheta = 1/2$ and shows the usual functional Gompertzian growth, in coherence with the behavior expected from the model.

References

Abdulle, A., Cohen, D., Vilmart, G., & Zygalakis, K. C. (2012). High weak order methods for stochastic differential equations based on modified equations. *SIAM Journal of Scientific Computing, 34*(3), a1800–a1823. https://doi.org/10.1137/110846609

Abdulle, A., Vilmart, G., & Zygalakis, K. C. (2013). Weak second order explicit stabilized methods for stiff stochastic differential equations. *SIAM Journal on Scientific Computing, 35*(4), A1792–A1814. Retrieved from https://doi.org/10.1137/12088954X.

Abdulle, A., Vilmart, G., & Zygalakis, K. C. (2014). High order numerical approximation of the invariant measure of ergodic sdes. *SIAM Journal on Numerical Analysis, 52*(4), 1600–1622. Retrieved from https://doi.org/10.1137/130935616.

Arnold, L. L. (1974). *Stochastic differential equations: Theory and applications*. New York: Wiley.

Bouchard, B., & Touzi, N. (2004). Discrete-time approximation and montecarlo simulation of backward stochastic differential equations. *Stochastic Processes and their Applications, 111*(2), 175–206. Retrieved from https://EconPapers.repec.org/RePEc:eee:spapps:v:111:y:2004:i:2:p:175-206.

Buckwar, E., & D'Ambrosio, R. (2021). Exponential mean-square stability properties of stochastic linear multistep methods. *Advances in Computational Mathematics, 47*(4), 14. (Id/No 55) https://doi.org/10.1007/s10444-021-09879-2.

Buckwar, E., Horváth-Bokor, R., & Winkler, R. (2005). *Asymptotic meansquare stability of two-step methods for stochastic ordinary differential equations*. Humboldt-Universität zu Berlin, Mathematisch-Naturwissenschaftliche Fakultät II, Institut für Mathematik. https://doi.org/10.18452/2597.

Buckwar, E., RöSSler, A., & Winkler, R. (2010). Stochastic runge–kutta methods for itô sodes with small noise. *SIAM Journal on Scientific Computing, 32*(4), 1789–1808. Retrieved from https://doi.org/10.1137/090763275.

Buckwar, E., & Sickenberger, T. (2011). A comparative linear meansquare stability analysis of maruyama- and milstein-type methods. *Mathematics and Computers in Simulation, 81*(6), 1110–1127. Retrieved from https://www.sciencedirect.com/science/article/pii/S0378475410003058, https://doi.org/10.1016/j.matcom.2010.09.015.

Burrage, K., & Burrage, P. M. (2012). Low rank runge–kutta methods, symplecticity and stochastic hamiltonian problems with additive noise. *Journal of Computational and Applied Mathematics, 236*(16), 3920–3930. Retrieved from https://www.sciencedirect.com/science/article/pii/S0377042712001240 (40 years of numerical analysis: "Is the discrete world an approximation of the continuous one or is it the other way around?") https://doi.org/10.1016/j.cam.2012.03.007.

Burrage, K., & Burrage, P. M. (2014). Structure-preserving runge-kutta methods for stochastic hamiltonian equations with additive noise. *Numerical Algorithms, 65*(3), 519–532. Retrieved from https://doi.org/10.1007/s11075-013-9796-6.

Burrage, K., & Tian, T. (2004). Implicit stochastic runge-kutta methods for stochastic differential equations. *BIT (Copenhagen), 44*(1), 21–39.

Chartier, P., Makazaga, J., Murua, A., & Vilmart, G. (2014). Multi-revolution composition methods for highly oscillatory differential equations. *Numerische Mathematik, 128*(1), 167–192. Retrieved from https://doi.org/10.1007/s00211-013-0602-0.

Chen, C., Cohen, D., D'Ambrosio, R., & Lang, A. (2020). Drift-preserving numerical integrators for stochastic hamiltonian systems. *Advances in Computational Mathematics, 46*(2). Retrieved from https://doi.org/10.1007/s10444-020-09771-5.

Chen, C., Cohen, D., & Hong, J. (2016). Conservative methods for stochastic differential equations with a conserved quantity. *International Journal of Numerical Analysis and Modeling, 13*(3), 435–456.

Cohen, D. (2012). On the numerical discretisation of stochastic oscillators. *Mathematics and Computers in Simulation, 82*(8), 1478–1495. Retrieved from https://www.sciencedirect.com/science/article/pii/S0378475412000286, https://doi.org/10.1016/j.matcom.2012.02.004.

Cohen, D., & Dujardin, G. (2014). Energy-preserving integrators for stochastic poisson systems. *Communications in Mathematical Sciences, 12*(8), 1523–1539. https://doi.org/10.4310/CMS.2014.v12.n8.a7

Cohen, D., & Vilmart, G. (2022). Drift-preserving numerical integrators for stochastic poisson systems. *International Journal of Computer Mathematics, 99*(1), 4–20. Retrieved from https://doi.org/10.1080/00207160.2021.1922679

de la Cruz, H. (2020). Stabilized explicit methods for the approximation of stochastic systems driven by small additive noises. *Chaos, Solitons & Fractals, 140*, 110195. Retrieved from https://www.sciencedirect.com/science/article/pii/S0960077920305919, https://doi.org/10.1016/j.chaos.2020.110195.

de la Cruz, H., Biskay, R., Jimenez, J., Carbonell, F., & Ozaki, T. (2010). High order local linearization methods: An approach for constructing a-stable explicit schemes for stochastic differential equations with additive noise. *BIT Numerical Mathematics, 50*(3), 509–539. Retrieved from https://doi.org/10.1007/s10543-010-0272-6.

de la Cruz, H., Jimenez, J., & Biscay, R. (2019). On the oscillatory behavior of coupled stochastic harmonic oscillators driven by random forces. *Statistics & Probability Letters, 146*(C), 85–89.

Retrieved from https://ideas.repec.org/a/eee/stapro/v146y2019icp85-89.html https://doi.org/10.1016/j.spl.2018.11.001.

de la Cruz, H., Jimenez, J., & Zubelli, J. P. (2017). Locally Linearized methods for the simulation of stochastic oscillators driven by random forces. *BIT Numerical Mathematics, 57*(1), 123–151. Retrieved from https://doi.org/10.1007/s10543-016-0620-2.

D. Higham, P. K. (2021). An introduction to the numerical simulation of stochastic differential equations. *SIAM,* xvi+277. Retrieved from https://doi.org/10.1365/s13291-021-00242-4.

Donnet, S., & Samson, A. (2013). A review on estimation of stochastic differential equations for pharmacokinetic/pharmacodynamic models. *Advanced Drug Delivery Reviews, 65*(7), 929–939. Retrieved from https://www.sciencedirect.com/science/article/pii/S0169409X13000501 (Mathematical modeling of systems pharmacogenomics towards personalized drug delivery) https://doi.org/10.1016/j.addr.2013.03.005.

D'Ambrosio, R., & Giovacchino, S. D. (2021a). Mean-square contractivity of stochastic q-methods. *Communications in Nonlinear Science and Numerical Simulation, 96*, 105671. Retrieved from https://www.sciencedirect.com/science/article/pii/S1007570420305013, https://doi.org/10.1016/j.cnsns.2020.105671.

D'Ambrosio, R., & Giovacchino, S. D. (2021). Nonlinear stability issues for stochastic runge-kutta methods. *Communications in Nonlinear Science and Numerical Simulation, 94*, 105549.

D'Ambrosio, R., & Scalone, C. (2021a). On the numerical structure preservation of nonlinear damped stochastic oscillators. *Numerical Algorithms, 86*(3), 933–952. Retrieved from https://doi.org/10.1007/s11075-020-00918-5.

D'Ambrosio, R., & Scalone, C. (2021b). Two-step runge-kutta methods for stochastic differential equations. *Applied Mathematics and Computation, 403*, 125930. Retrieved from https://www.sciencedirect.com/science/article/pii/S0096300320308833, https://doi.org/10.1016/j.amc.2020.125930.

Fang, W., & Giles, M. B. (2020). Adaptive Euler–Maruyama method for SDEs with nonglobally Lipschitz drift. *The Annals of Applied Probability, 30*(2), 526–560. Retrieved from https://doi.org/10.1214/19-AAP1507

Ferrante, L., Bompadre, S., Leone, L., & M.P., M. (2005). A stochastic formulation of the gompertzian growth model for in vitro bactericidal kinetics: parameter estimation and extinction probability. *Biometrical journal. Biometrische Zeitschrift, 470*(2), 309–318. Retrieved from https://pubmed.ncbi.nlm.nih.gov/16053255/, https://doi.org/10.1002/bimj.200410125.

Gardiner, C. W. (2004). *Handbook of stochastic methods, for physics, chemistry and the natural sciences*. Springer. https://link.springer.com/book/9783540707127

Gobet, E. (2022). *Monte-carlo methods and stochastic processes: From linear to non-linear* (1st ed.). Chapman and Hall/CRC. https://doi.org/10.1201/9781315368757.

Gobet, E., Lemor, J.-P., & Warin, X. (2005). A regression-based Monte Carlo method to solve backward stochastic differential equations. *The Annals of Applied Probability, 15*(3), 2172–2202. Retrieved from https://doi.org/10.1214/105051605000000412.

Higham, D. J. (2000). Mean-square and asymptotic stability of the stochastic theta method. *SIAM Journal on Numerical Analysis, 38*(3), 753-769. Retrieved from https://doi.org/10.1137/S003614299834736X.

Higham., D. J. (2001). An algorithmic introduction to numerical simulation of stochastic differential equations. *SIAM Review, 43*(3), 525–546. Retrieved from https://doi.org/10.1137/S0036144500378302.

Higham, D. J., & Kloeden, P. (2005). Numerical methods for nonlinear stochastic differential equations with jumps. *Numerische Mathematik, 101*(1), 101–119. Retrieved from https://doi.org/10.1007/s00211-005-0611-8.

Hutzenthaler, M., & Jentzen, A. (2015). Numerical approximations of stochastic differential equations with non-globally lipschitz continuous coefficients. *Memoirs of the American Mathematical Society*. https://doi.org/10.1090/memo/1112

Hutzenthaler, M., Jentzen, A., & Kloeden, P. E. (2011). Strong and weak divergence in finite time of euler's method for stochastic differential equations with non-globally lipschitz continuous coefficients. *Proceedings of the Royal Society A: Mathematical, Physical and Engineering Sciences, 467*(2130), 1563–1576. Retrieved from https://royalsocietypublishing.org/doi/abs/10.1098/rspa.2010.0348, https://doi.org/10.1098/rspa.2010.0348.

Jimenez, H., & de la Cruz Cancino, J. C. (2012). Convergence rate of strong local linearization schemes for stochastic differential equations with additive noise. *BIT Numerical Mathematics, 52*(2), 357–382. Retrieved from https://doi.org/10.1007/s10543-011-0360-2.

Kloeden, P. E. (2002). The systematic derivation of higher order numerical schemes for stochastic differential equations. *Milan Journal of Mathematics, 70*(1), 187–207. Retrieved from https://doi.org/10.1007/s00032-002-0006-6

Kloeden, P. E., & Platen, E. (1992). *Numerical solution of stochastic differential equations*. Berlin, Heidelberg: Springer. https://doi.org/10.1007/978-3-662-12616-5.

Ma, J., & Zhang, J. (2002). Path regularity for solutions of backward stochastic differential equations. *Probability Theory and Related Fields, 1222*(2), 163–190. Retrieved from https://doi.org/10.1007/s004400100144.

Ma, Q., Ding, D., & Ding, X. (2012). Symplectic conditions and stochastic generating functions of stochastic runge–kutta methods for stochastic hamiltonian systems with multiplicative noise. *Applied Mathematics and Computation, 219*(2), 635–643. Retrieved from https://www.sciencedirect.com/science/article/pii/S0096300312006613, https://doi.org/10.1016/j.amc.2012.06.053

Mao, X. (2007). *Stochastic differential equations and applications*. Chichester: Horwood.

Melbø, A. H., & Higham, D. J. (2004). Numerical simulation of a linear stochastic oscillator with additive noise. *Applied Numerical Mathematics, 51*(1), 89–99. Retrieved from https://www.sciencedirect.com/science/article/pii/S0168927404000285 https://doi.org/10.1016/j.apnum.2004.02.003.

Milstein, G. N. (1994). *Numerical integration of stochastic differential equations. Translation from the Russian* (Vol. 313). Dordrecht: Kluwer Academic Publishers.

Milstein, G. N., Repin, Y. M., & Tretyakov, M. V. (2002). Numerical methods for stochastic systems preserving symplectic structure. *SIAM Journal on Numerical Analysis, 40*(4), 1583–1604. https://doi.org/10.1137/S0036142901395588

Misawa, T. (2000). Energy conservative stochastic difference scheme for stochastic Hamilton dynamical systems. *Japan Journal of Industrial and Applied Mathematics, 17*(1), 119–128. https://doi.org/10.1007/BF03167340

Neuenkirch, A., Szölgyenyi, M., & Szpruch, L. (2019). An adaptive Euler- Maruyama scheme for stochastic differential equations with discontinuous drift and its convergence analysis. *SIAM Journal of Numerical Analysis, 57*(1), 378–403. https://doi.org/10.1137/18M1170017

Pardoux, E., & Peng, S. G. (1990). Adapted solution of a backward stochastic differential equation. *Systems and Control Letters, 14*(1), 55–61. https://doi.org/10.1016/0167-6911(90)90082-6

Rössler, A. (2006). Runge-Kutta methods for Itô stochastic differential equations with scalar noise. *BIT, 46*(1), 97–110. https://doi.org/10.1007/s10543-005-0039-7

Rössler, A. (2009). Second order Runge-Kutta methods for Itô stochastic differential equations. *SIAM Journal of Numerical Analysis, 47*(3), 1713–1738. https://doi.org/10.1137/060673308

Rössler, A. (2010). Runge-Kutta methods for the strong approximation of solutions of stochastic differential equations. *SIAM Journal of Numerical Analysis, 48*(3), 922–952. https://doi.org/10.1137/09076636X

Ruemelin, W. (1982). Numerical treatment of stochastic differential equations. *SIAM Journal of Numerical Analysis, 19*, 604–613. https://doi.org/10.1137/0719041

Saito, Y., & Mitsui, T. (1996). Stability analysis of numerical schemes for stochastic differential equations. *SIAM Journal of Numerical Analysis, 33*(6), 2254–2267. https://doi.org/10.1137/S0036142992228409.

Shreve, S. E. (2004). *Stochastic calculus for finance. II: Continuous-time models*. New York, NY: Springer.

Vom Scheidt, J. (1989). T. C. Gard (Ed.), *Introduction to Stochastic Differential Equations, 1988* (Vol. XI, 234, p. $ 78). New York-Basel, Marcel Dekker Inc. ISBN 0-8247-7776-X (Pure and Applied Mathematics 114). *Zeitschrift Angewandte Mathematik und Mechanik*, 69(8), 258–258. https://doi.org/10.1002/zamm.19890690808.

Dimitri Breda Associate Professor of numerical analysis at the Department of Mathematics, Computer Science and Physics, University of Udine, where he founded and leads the CDLab—Computational Dynamics Laboratory. His research interests are in the numerical and applied mathematical analysis of infinite-dimensional dynamical systems from population dynamics and control engineering.

Jung Kyu Canci Senior lecturer and researcher at University of Basel and of Applied Science in Lucerne. His research is in pure mathematics, Number Theory with particular interests in Arithmetic of Dynamical Systems, and in applied mathematics, Stochastic Processes in Finance. He is also the founder of several companies.

Raffaele D'Ambrosio Professor of Numerical Analysis at the University of L'Aquila, in Italy. He has been Fulbright Research Scholar in the Academic Year 2014–15 at Georgia Institute of Technology. His main research interests regard the numerical approximation of deterministic and stochastic evolutive problem and their geometric numerical integration.

Open Access This chapter is licensed under the terms of the Creative Commons Attribution 4.0 International License (http://creativecommons.org/licenses/by/4.0/), which permits use, sharing, adaptation, distribution and reproduction in any medium or format, as long as you give appropriate credit to the original author(s) and the source, provide a link to the Creative Commons license and indicate if changes were made.

The images or other third party material in this chapter are included in the chapter's Creative Commons license, unless indicated otherwise in a credit line to the material. If material is not included in the chapter's Creative Commons license and your intended use is not permitted by statutory regulation or exceeds the permitted use, you will need to obtain permission directly from the copyright holder.

Life Events that Cascade: An Excursion into DALY Computations

Young Lee, Thanh Vinh Vo, Derek Ni, and Gang Mu

1 Introduction

Background and motivation of the study. In most applications concerning point processes, we are generally confronted with the problem that we do not have the times at which events have taken place. In such cases, the only information we have is the number of events over a given interval, but the event times are unknown. For example, since there can be many automobile accidents per day in the USA, it is common to record the number of accidents for that day, rather than the exact times at which these events happened.

The purpose of the present article is to spell out closed-form solutions for some functionals of the *number of events* over a given *interval* within a class of point processes that exhibit *self* and *externally* excitatory interactions. Intuitively, a process is self-exciting if the occurrence of past events makes the occurrence of future events more probable.

The origin of point processes incorporating *both* self- and external excitations date back to 2002, when the authors Brémaud and Massoulié (2002) introduced this model under very general conditions. In this paper, we perform inference on a model that combines both self-exciting behavior and externally exciting components as

---

Y. Lee (✉) · D. Ni
F. Hoffmann-La Roche AG, Basel, Switzerland
e-mail: young.lee.yl2@roche.com

D. Ni
e-mail: derek.ni@roche.com

T. V. Vo
National University of Singapore, Queenstown, Singapore
e-mail: votv@comp.nus.edu.sg

G. Mu
University of Zurich, Zürich, Switzerland
e-mail: gang.mu@math.uzh.ch

© The Author(s) 2023
J. K. Canci et al. (eds.), *Quantitative Models in Life Science Business*,
SpringerBriefs in Economics, https://doi.org/10.1007/978-3-031-11814-2_7

set forth in Dassios and Zhao (2011), where they incorporated the shot-noise Cox processes (Mohler 2013) as the external component while maintaining the Hawkes process (Hawkes 1971) for self-excitations. We shall henceforth call this process the *Hawkes–Cox* model.

Related work. Inference for point processes with the combined elements of endogenous and exogenous components have been studied in the machine learning community. Linderman and Adams (2014) introduced a multidimensional counting process combining a sparse log Gaussian Cox process (Møller et al. 1998) and a Hawkes component to uncover latent networks in the point process data. They showed how the superposition theorem of point processes enables the formulation of a fully Bayesian inference algorithm. Mohler (2013) considered a self-exciting process with background rate driven by a log Gaussian Cox process and performed inference based on an efficient Metropolis adjusted Langevin algorithm for filtering the intensity. Simma and Jordan (2010) proposed an expectation–maximization inference algorithm for Cox-type processes incorporating self-excitations via marked point processes and applied to a very large social network data.

We note that a variety of methods have been developed for the estimation of self *or* external excitations for point processes: variational flavors (Mangion et al. 2011), expectation propagation (Cseke and Heskes 2011), and the usage of thinning points and uniformization for non-stationary renewal processes, (Gunter et al. 2014; Teh and Rao 2011).

Perhaps the most intimately related work to ours is that of Da Fonseca and Zaatour (2014) where the authors looked at the functionals of event counts over an interval for a Hawkes process (Hawkes 1971). In this work, we proposed an inference procedure for the Hawkes–Cox model which captures both the self and external excitatory relationships over a given interval. As a by product, we extend some theoretical calculations needed to give explicit higher moments to calculate the covariance structure while generalizing the results of Da Fonseca and Zaatour (2014).

Contributions. Consider the scenario where we do not know the exact times at which events happen, but we have the number of events that have occurred in a given interval. We propose an inference procedure to learn the Hawkes–Cox model that can capture both self- and external excitatory relationships under these circumstances. Our major contributions are as follows: (*i*) We develop closed-form solutions of some functionals for the Hawkes–Cox process. (*ii*) Special attention is given to the covariance function of event counts over an interval. We show that the covariance structure over the number of jumps in an interval enjoys analytical tractability. This measure is extremely important as it supports the clustering property of the self-exciting property of the Hawkes process and the external excitation features of the Cox process. (*iii*) We developed an inference procedure through L_2 regularization to learn parameters where estimation is almost instantaneous. (*iv*) We finally conclude by demonstrating the usefulness of applying Hawkes–Cox to model financial trade volumes.

2 The Hawkes–Cox Framework

Background on point processes. This section introduces some pieces of counting process theory needed in what follows. The *homogeneous Poisson process* $\hat{N} = \{\hat{N}(t) : t \geq 0\}$ with intensity m has the following properties: (a) The process starts at 0 with $\hat{N} = 0$; (b) the process has independent increments, i.e., for any $t_i, i = 0, ..., n$ the increments $\hat{N}_{t_i} - \hat{N}_{t_{i-1}}$ are mutually independent; (c) there exists a non-decreasing right continuous function $m : [0, \infty) \to [0, \infty)$ with $m(0) = 0$ such that the increments $\hat{N}_t - \hat{N}_s$ for $0 < s < t < \infty$ have a Poisson distribution with mean value function $m(t) - m(s)$.

The Hawkes–Cox Model. We are interested in a counting process $N(t)$ whose behavior is affected by past events, which contains both self- and externally excitation elements. Our point process $N(t)$ has a non-negative $\mathcal{F}_t$– stochastic intensity function $\lambda(t)$ of the form:

$$\lambda(t) = \mathcal{B}_0(t) + \sum_{i:t>T_i} \mathcal{H}(Y_i, t - T_i) + \sum_{i:t>S_i} \mathcal{C}(X_i, t - S_i), \qquad (1)$$

where $\mathcal{B}, \mathcal{H}$ and $\mathcal{C}$ are functions whose definitions will be made precise in Definition 1 below. The sequence $(T_i)_{i \geq 1}$ denotes the event times of N, where the occurrence of an event induces the intensity to grow by an amount $\mathcal{H}(Y_i, t - T_i)$: this element captures self-excitation. At the same time, external events can occur at times S_i and stimulate with a portion of $\mathcal{C}(X_i, t - T_i)$: this is the externally excited part. The quantities X and Y are positive random elements describing the amplitudes by which λ increases during event times. The quantity $\mathcal{B}_0 : \mathbb{R}_+ \mapsto \mathbb{R}_+$ denotes the *deterministic* base intensity. We write $N_t := N(t)$ and $\lambda_t := \lambda(t)$ to ease notation and $\{\mathcal{F}_t\}$ being the history of the process and contains the list of times of events up to and including t, i.e. $\{T_1, T_2, ..., T_{N_t}\}$.

We now give specific forms for $\mathcal{B}, \mathcal{H}$ and $\mathcal{C}$ through the definition of the Hawkes–Cox model.

Definition 1 The Hawkes–Cox process is a point process N on $\mathbb{R}^+$ with the non-negative $\mathcal{F}_t$ conditional random intensity

$$\lambda_t = a + (\lambda_0 - a)e^{-\delta t} + \sum_{i=1}^{N_t} Y_i e^{-\delta(t - T_i)} + \sum_{i=1}^{J_t} X_i e^{-\delta(t - S_i)}, \qquad (2)$$

for $t \geq 0$, where we have the following features:

- **Deterministic background.** $a \geq 0$ is the constant mean-reverting level, $\lambda_0 > a$ is the initial intensity at time $t = 0$, $\delta > 0$ is the constant rate of exponential decay. $\mathcal{B}_0(t) = a + (\lambda_0 - a)e^{-\delta t}$;
- **External excitations.** X_i are levels of excitation from an external factor. They form a sequence of independent and identically distributed positive elements with

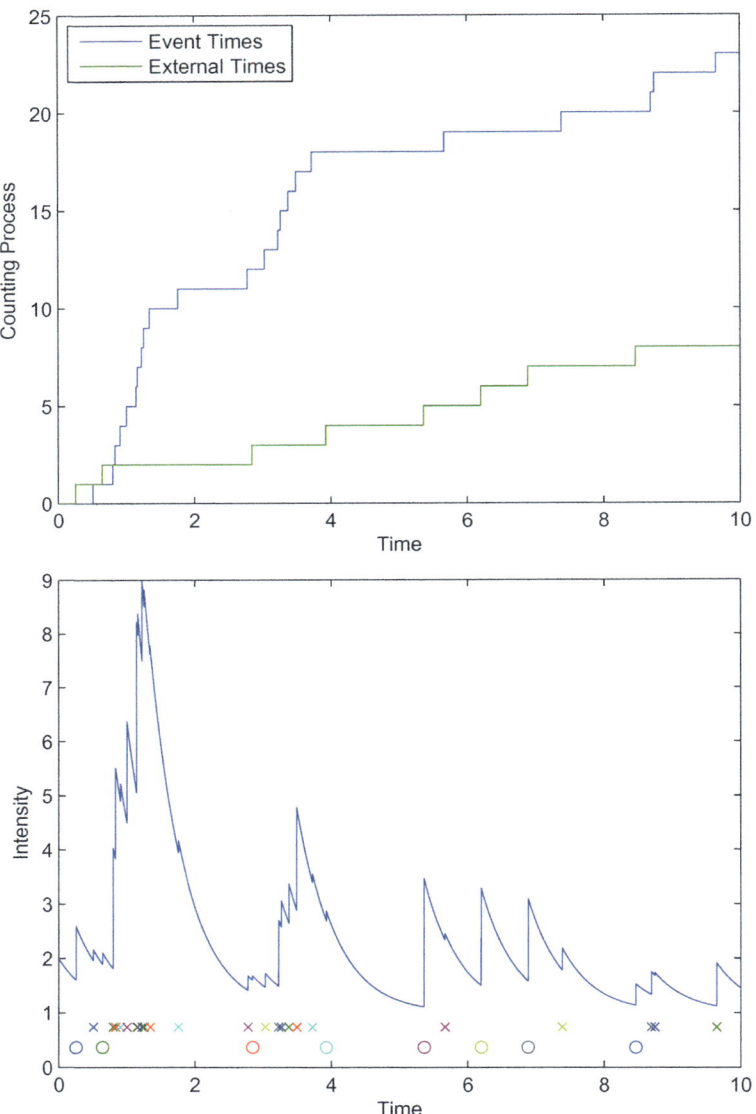

Fig. 1 A sample of a Hawkes–Cox process. In the top plot, the blue line corresponds to the counting process of interested event times, while the black line shows the counting process of the external event times. The bottom graph plots the realized intensity function λ_t

distribution function $H(c)$, $c > 0$. S_i are the times at which external events happen and it follows a Poisson process J_t of constant rate $\rho > 0$. Similar kernel types are used in Fisher and Banerjee (2010); Du et al. (2015), to name a few. Note that $\mathcal{C}(X_i, t - S_i) := X_i e^{-\delta(t-S_i)}$.

- **Self-excitations.** Y_i are levels of *self-excitation*, a sequence of independent and identically distributed positive elements with distribution function $G(h)$, $h > 0$, occurring at random T_i. Following the occurrence of events, the impact of these events will saturate and the rate at which this occurs is determined by the constant δ. Note that $\mathcal{H}(Y_i, t - T_i) := Y_i e^{-\delta(t-T_i)}$.

For illustration, we present a sample simulation path obtained from the above parameterization in Fig. 1, showing generated event times. From this figure, we can see that the intensity process is excited by both the event times of interest (in crosses ×) and the external events (in circle ○).

Note that from Definition 1, if we set $X \equiv 0$ and Y to be a constant, we retrieve the model proposed by Hawkes (1971). If we set $Y \equiv 0$ and X to be a positive random elements, we get the model proposed by Cox and Isham (1980). In addition, setting $X = Y = 0$ returns us the inhomogeneous Poisson process, Daley and Vere-Jones (2003). Furthermore, letting $X = Y = 0$ and $\lambda_0 = a$ simplify to the Poisson process.

2.1 The Choice of Kernel and the Rôle of δ

The kernel for the externally excited component is known as the shot noise (Cox and Isham 1980; Møller 2003) where $\mathcal{C}(X_i, t - S_i) = X_i e^{-\delta(t-S_i)}$. This is a deliberate choice with δ being shared between the background rate $\mathcal{E}_0$ $\mathcal{H}$ and $\mathcal{C}$ to ensure that the process inherits the Markov property, see Brémaud and Massoulié (2002); Blundell et al. (2012). This property is essential to the derivation of the functionals in Sect. 3. The term δ determines the rate at which the process decays exponentially from following arrivals of self-excited and externally excited events.

3 Dynkin's Formula

The Markov property is the *key property* that allows one to invoke certain tools to obtain the moments for the number of events in *an interval*, rather than the number of events at a current time $t > 0$. Among these tools are the *infinitesimal generator* and martingale techniques. For a Markov process X_t, consider a function $f : D \to \mathbb{R}$. The infinitesimal generator of the process denoted $\mathcal{A}$, is defined by

$$\mathcal{A}f(x) = \lim_{h \to 0} \frac{\mathbb{E}\left[f(X_{t+h})|X_t = x\right] - f(x)}{h}.$$

For every function f in the domain of the infinitesimal generator, the process

$$M_t := f(X_t) - f(X_0) - \int_0^t \mathcal{A}f(X_u)du$$

is a martingale (Øksendal and Sulem 2007). Thus, for $t > s$, we have

$$\mathbb{E}\left[f(X_t) - \int_0^s \mathcal{A}f(X_u)du\Big|\mathcal{F}_s\right] = f(X_s) - \int_0^s \mathcal{A}f(X_u)du$$

by the martingale property of M. Rearranging, we finally obtain Dynkin's formula (see, Øksendal and Sulem (2007))

$$\mathbb{E}[f(X_t)|\mathcal{F}_s] = f(X_s) + \mathbb{E}\left[\int_s^t \mathcal{A}f(X_u)du\Big|\mathcal{F}_s\right]. \tag{3}$$

In the following section, we will heavily rely on this formula to compute some distributional properties of the Hawkes–Cox process.

4 Theoretical Moments

Our aim is to compute the following quantities: mean, variance and covariance of the *number of events* in an *interval* τ, rather than at a fixed time t, as investigated by Dassios and Zhao (2011). To do so, we appeal to the techniques and manipulations presented in the previous section.

To ease notations, the first and second moments of levels of self- and external excitations X and Y, respectively, are denoted by

$$\mu_{1G} := \int_0^\infty h dG(h), \quad \mu_{2G} := \int_0^\infty h^2 dG(h),$$

$$\mu_{1H} := \int_0^\infty c dH(c), \quad \mu_{2H} := \int_0^\infty c^2 dH(c)$$

as well as the constant

$$k := \delta - \mu_{1G}.$$

First note that the joint process of $\{(\lambda_t, N_t)\}_{t \geq 0}$ is a Markov process. This can be seen by proving that the expression λ_{T_k} only depends on $\lambda_{T_{k-1}}$ and $\{N_t : T_{k-1} \leq t \leq T_k\}$. By the Markov property and using the results in Øksendal and Sulem (2007); Davis (1984), the infinitesimal generator of the process (t, λ_t, N_t) acting on a function $f(t, \lambda, n)$ is given by

$$\mathcal{A}f(t,\lambda,n) = -\delta(\lambda-a)\frac{df}{d\lambda} + \lambda\left(\int_0^\infty f(t,\lambda+y,n+1)dG(y) - f(t,\lambda,n)\right)$$
$$+ \rho\left(\int_0^\infty f(t,\lambda+c,n)dH(c) - f(t,\lambda,n)\right).$$

4.1 The Moments of Counts in an Interval

In order to obtain the expected number of jumps for an interval $(s,t]$, we apply Dynkin's formula as in Eq. (3). By setting $f(t,\lambda,n) = n$, we have that $\mathcal{A}n = \lambda$. Since $N_t - N_s - \int_s^t \lambda_u du$ is a martingale, we have $\mathbb{E}[N_t - N_s - \int_s^t \mathcal{A}N_u du | \mathcal{F}_s] = 0$. Rearranging, we get

$$\mathbb{E}[N_t|\mathcal{F}_s] = N_s + \mathbb{E}\left[\int_s^t \lambda_u du\right]. \tag{4}$$

Due to Fubini's Theorem, we interchange the order of expectation and the integral subsequently yields

$$\mathbb{E}[N_t|\mathcal{F}_s] = N_s + \int_s^t \mathbb{E}[\lambda_u|\mathcal{F}_s] du. \tag{5}$$

Remark that this equation could have been obtained by recalling that $(N_t - \int_0^t \lambda_s ds)_{t \in [0,T]}$ is a martingale, by definition of the intensity of a point process, Daley and Vere-Jones (2003).

We can now give an explicit expression for the expectation of the number of events in an interval.

Proposition 1 *The expectation of the number of jumps over an interval τ is given by*

$$\lim_{t \to \infty} \mathbb{E}\left[N_{t+\tau} - N_t | \lambda_0\right] = \frac{a\delta}{\delta - \mu_{1G}}\tau =: h_1^\tau \tag{6}$$

Proof An application of tower property of expectation yields

$$\mathbb{E}[N_t - N_s | \lambda_0] = \mathbb{E}[\mathbb{E}[N_t - N_s | \mathcal{F}_s] | \mathcal{F}_0]$$
$$= \mu_1(t-s) + \frac{1}{k}(1 - e^{-k(t-s)})(\mathbb{E}[\lambda_s|\mathcal{F}_0] - \mu_1)$$
$$= \mu_1(t-s) - \frac{1}{k}(\lambda_0 - \mu_1)(e^{-kt} - e^{ks}),$$

for $s < t$. Setting $t \leftarrow t+\tau$ and $s \leftarrow t$ and letting $t \to \infty$ and under the stationary condition $\mu_{1G} < \delta$, we obtain the expression in Eq. (6). ∎

We interpret this quantity as the long-run expectation of the number of events during a given time interval of length τ.

We now calculate the long-run variance of the number of jumps during an interval of length τ. We first note that

$$\text{Var}[N_t - N_s | \mathcal{F}_0]$$
$$= \mathbb{E}[(N_t - N_s)^2 | \mathcal{F}_0] - (\mathbb{E}[N_t - N_s | \mathcal{F}_0])^2. \tag{7}$$

We proceed by calculating the first quantity on the RHS in Eq. (7). First, note that

$$\mathbb{E}[(N_t - N_s)^2 | \mathcal{F}_0] = \mathbb{E}[\mathbb{E}[(N_t - N_s)^2 | \mathcal{F}_s] | \mathcal{F}_0].$$

Given the previous two results, we can now state the expression for the variance of the number of events in an interval of length τ:

Proposition 2 *The variance of the number of jumps over an interval τ is given by*

$$\lim_{t \to \infty} \text{Var}(N_{t+\tau} - N_t | \mathcal{F}_0)$$
$$= \lim_{t \to \infty} \mathbb{E}[(N_{t+\tau} - N_t)^2 | \mathcal{F}_0] - \lim_{t \to \infty} (\mathbb{E}[N_{t+\tau} - N_t | \mathcal{F}_0])^2$$
$$= \frac{\theta}{k} + \frac{2}{k}\left(\mu_{1G}\mu_1 - \mu_1^2 + \frac{(2\theta + \mu_{2G})\theta}{2k^2} + \frac{\rho\mu_{2H}}{2k}\right)\tau - \frac{2}{k^2}\left(\frac{(2\theta + \mu_{2G})\theta}{2k^2} + \frac{\rho\mu_{2H}}{2k} + \mu_{1G}\mu_1 + \mu_1^2\right)$$
$$:= h_2^\tau. \tag{8}$$

4.2 The Covariance

We now turn to perhaps the most important quantity: the covariance function which carries information regarding the clustering nature of Hawkes–Cox. To calculate this cross-expectation of the number of events during different time intervals, we first need to determine the following expression:

$$\mathbb{E}[(N_{t_1} - N_t)(N_{t_3} - N_{t_2}) | \mathcal{F}_0] - \mathbb{E}[(N_{t_1} - N_t) | \mathcal{F}_0]\mathbb{E}[(N_{t_3} - N_{t_2}) | \mathcal{F}_0],$$

where $t_0 < t < t_1 < t_2 < t_3$. For simplicity, we let $\tau := t_1 - t$ and $\tau = t_3 - t_2$. We further define the lag of Hawkes–Cox as $\tilde{\tau} := t_2 - t_1$. We state the following:

Proposition 3 *The covariance of the number of jumps in a given interval τ with lag $\tilde{\tau}$ can be explicitly expressed as*

$$\text{Cov}(\tau, \tilde{\tau}) = \lim_{t \to \infty} \left(\mathbb{E}[(N_{t+\tau} - N_t)(N_{t+2\tau+\tilde{\tau}} - N_{t+\tau+\tilde{\tau}})] - \mathbb{E}[(N_{t+\tau} - N_t)]\mathbb{E}[(N_{t+2\tau+\tilde{\tau}} - N_{t+\tau+\tilde{\tau}})]\right)$$
$$= (\mu_1\tau)^2 + \left[\left(\frac{1 - e^{-k\tau}}{k}\right)^2 e^{-k\tilde{\tau}} \left(\frac{(2\theta + \mu_{2G})\theta}{2k^2} + \frac{\rho\mu_{2H}}{2k} + \mu_1\mu_{1G} - \mu_1^2\right)\right] =: h_3^\tau. \tag{9}$$

5 Learning and Optimization

We present an optimization technique to learn the parameters of the Hawkes–Cox process. Consider the cost function $\mathcal{J}$ of the following form:

$$\mathcal{J}(\varphi) = \sum_{i=1}^{3} \left| \lim_{t \to \infty} C_i^{\varphi}(t) - C_i \right|^2 + \lambda_{reg} F(\varphi),$$

where $C_i^{\varphi}(t)$ denotes the theoretical function of moments for Hawkes–Cox. $C_1^{\varphi}(t) = \mathbb{E}(N_{t+\tau} - N_t)$, $C_2^{\varphi}(t) = \text{Var}(N_{t+\tau} - N_t)$ and

$$C_3^{\varphi}(t) = \mathbb{E}[(N_{t+\tau} - N_t)(N_{t+2\tau+\tilde{\tau}} - N_{t+\tau+\tilde{\tau}})] - \mathbb{E}[(N_{t+\tau} - N_t)]\mathbb{E}[(N_{t+2\tau+\tilde{\tau}} - N_{t+\tau+\tilde{\tau}})]$$

and C_i being the corresponding *empirical* function of moments in a given interval τ and lag $\tilde{\tau}$ where λ_{reg} is a parameter that controls the level of regularization in the optimization scheme with F being the regularization term. The first role of the penalization term F is to ensure uniqueness of the solution. Indeed, if F is convex in the φ, then for λ_{reg} large enough, the entire functional $\mathcal{J}$ will be convex and the solution will be unique. Note that the difference in the square terms are deviations of the theoretical functions of moments against the corresponding empirical values. Let the parameters of interest be φ. The learned parameter $\hat{\varphi}$ is obtained from

$$\hat{\varphi} = \arg\min_{\varphi} \mathcal{J}(\varphi),$$

where φ is the parameter of interest. The details of optimization tools are presented in the experiment section.

6 Synthetic Experiments

We evaluate our proposed optimization method using both simulated and real-world data, and show that our approach significantly outperforms the baseline.

With estimation being almost instantaneous, we demonstrate the usefulness of Hawkes–Cox to model financial trade volumes. Empirical evidence have shown that trading activity is not a random process but rather produces time sequences that exhibit clustering behavior (Filimonov and Sornette 2012). If it was the case, a homogeneous Poisson process would have been a suitable candidate to model trade arrival times. Possible exogenous liquidity and news shocks are being capture through the external excitation component of Hawkes–Cox.

We show that our model achieves better accuracy in the prediction compared to MLE and also a homogeneous Point process (HPP).

6.1 Calibration

The experiments on assessing the stability and accuracy of the learned parameters from the Hawkes–Cox process are studied. First, we illustrate the generation of the synthetic data. Then we briefly review the MLE method before presenting the calibration results obtained with the proposed optimization method.

6.1.1 Synthetic Data Generation

For simplicity, we assume the levels of excitation X and Y be drawn from random elements of exponential distribution with mean parameter ψ. One reason is that this makes the data generation process simple since an efficient simulation algorithm is available (Dassios and Zhao 2011). To avoid giving more weight to some parameters over others, we simply assume the following ground truth parameters for the simulation of the synthetic data. All parameters are set to a value of 1 except for the decay δ and the initial intensity λ_0. The decay parameter δ is set to a value of 2 because it has to be greater than ψ to respect the stationarity condition (Brémaud and Massoulié 1996; Brémaud et al. 2002), while λ_0 is set to 2 such that $(\lambda_0 - a)$ is non-zero. In addition, we arbitrarily set the maturity time T to 5,000 to allow for long-term stationary conditions to be achieved. We spell out the values of the parameters in Table 2 for ease of reading.

6.1.2 Maximum Likelihood Estimation

For practical applications, the MLE method has frequently been employed as the standard for parameter estimation on statistical models due to the lack of efficient and fast inference algorithms for large datasets. Presumably, the popularity of using MLE is due to Ozaki (1979) who first applied MLE procedures to estimate parameters of a Hawkes process. For example, despite the Markov chain Monte Carlo (MCMC) methods giving exact solutions in the limiting cases, they may be too slow for many practical applications (Kuss and Rasmussen 2005). Other inference techniques, albeit faster than MCMC, are still comparatively slower compared to MLE for practical purposes. Here, we use MLE to learn the set of parameters $\varphi = \{a, \delta, \psi, \rho\}$.

To account for randomness from the simulations, the MLE result is obtained from 100 samples of Hawkes–Cox processes simulated with the ground truth parameters described in Sect. 6.1.1 or Table 2. The technical approach for computing the minimum is by using a *bounded constrained optimizer* built upon MATLAB function *fminsearch*, which in turn employs the Nelder–Mead simplex method (Nelder and Mead 1965). We bound our search space from 0 to 10.

We find that the learned MLE parameters, presented in the first row of Table 3, are close to the ground truth parameters, where their standard errors are displayed in the corresponding brackets. However, a careful inspection of their 95 % confidence intervals (i.e., ± two standard errors from the estimates) suggests that the estimates are not adequate, as the confidence intervals for a, δ and ψ failed to contain the true values. Additionally, in Table 3, we compute the mean absolute difference (MAD) of the learned parameters against their ground truth values. This MAD value for MLE suggests that the learned parameters only deviate from (on average) the true values by 0.13.

We report that the MLE method takes quadratic time to run due to the evaluation of the loglikelihood function. In our case, we find that each MLE took around 846 s to complete.

6.1.3 L_2 – Regularization

We replicate the above experiments but now with the proposed L_2–regularization method described Sect. 5. Since estimation is almost instantaneous, we perform the assessment by simulating 1,000 sample paths of Hawkes–Cox processes. We repeat the experiments for different configurations of the tuning parameters, that is, interval τ in the range of 0.5–10.0, and regularization parameter λ_{reg} in the range of 0.0–1.0.

We present the MAD of the learned parameters against the true values corresponding to various configurations in Table 1, and the most accurately learned parameters from varying the λ_{reg} is tabulated in Table 3. Interestingly, we find that the higher

Table 1 The mean absolute differences for the learned parameters against their true values, obtained with our inference technique over various configurations of τ and λ_{reg}. The MADs are computed by averaging over 1,000 simulation paths. The results, for which the lower MAD the better, that are lower than 0.01 are highlighted in bold. Observe that as the interval τ increases, the stronger the regularization parameter λ_{reg} required for better parameter estimations

INTERVAL	REGULARIZATION PARAMETER λ_{reg}										
τ	0.0	0.1	0.2	0.3	0.4	0.5	0.6	0.7	0.8	0.9	1.0
1.0	**0.0014**	0.1835	0.3369	0.7770	0.9804	1.0033	1.0238	1.0422	1.0592	1.0748	1.0900
2.0	**0.0096**	0.0190	0.0420	0.0593	0.0779	0.0928	0.1046	0.1197	0.1309	0.1416	0.1503
3.0	0.0105	**0.0098**	0.0115	0.0217	0.0357	0.0358	0.0459	0.0550	0.0593	0.0694	0.0721
4.0	0.0440	0.0186	0.0115	**0.0043**	**0.0086**	0.0200	0.0261	0.0323	0.0313	0.0449	0.0534
5.0	0.0715	0.0537	0.0253	0.0224	0.0202	**0.0058**	0.0163	0.0119	0.0173	0.0185	0.0328
6.0	0.1037	0.1041	0.0813	0.0610	0.0318	0.0418	0.0195	0.0217	**0.0093**	**0.0092**	0.0105
7.0	0.1200	0.1985	0.1225	0.1173	0.0789	0.0476	0.0376	0.0409	0.0278	0.0181	0.0150
8.0	0.1445	0.2668	0.1958	0.1685	0.1269	0.0941	0.0932	0.0561	0.0370	0.0430	0.0199

Table 2 Ground truth parameters used in simulating the synthetic data

NAME	MATURITY TIME	INITIAL INTENSITY	BACKGROUND	DECAY	EXCITATION	EXTERNAL INTENSITY
VALUE	$T = 5000$	$\lambda_0 = 2$	$a = 1$	$\delta = 2$	$\psi = 1$	$\rho = 1$

Table 3 The learned parameters $\{a, \delta, \psi, \rho\}$ and their standard errors (enclosed in brackets) associated with the MLE method and the proposed inference technique. For the proposed optimization method, the results with the best MAD correspond to each τ are shown. We note that, with the exception of the MLE, the confidence intervals of the learned estimates contain the true parameters

Method	τ	λ_{reg}	a (true = 1)		δ (true = 2)		ψ (true = 1)		ρ (true = 1)		MAD
MLE	–	–	0.8785	0.0343	2.2155	0.0707	1.1778	0.0320	1.0011	0.0141	0.1290
Functional Moment Matching Method	1.0	0.0	**1.0000**	**0.0483**	**2.0046**	**0.1217**	**1.0006**	**0.0515**	**1.0003**	**0.0173**	**0.0014**
	2.0	0.0	1.0044	0.0498	2.0239	0.1614	1.0092	0.0703	0.9992	0.0200	0.0096
	3.0	0.1	1.0090	0.0560	2.0227	0.2440	1.0053	0.1062	0.9980	0.0219	0.0098
	4.0	0.3	1.0047	0.0594	1.9955	0.3290	0.9955	0.1493	0.9963	0.0249	0.0043
	5.0	0.5	1.0028	0.0641	1.9927	0.4526	0.9933	0.2074	0.9935	0.0259	0.0058
	6.0	0.9	1.0013	0.0668	1.9817	0.5067	0.9893	0.2369	0.9934	0.0265	0.0092
	7.0	1.0	0.9991	0.0719	2.0373	0.6516	1.0160	0.3063	0.9943	0.0306	0.0150
	8.0	1.0	1.0004	0.0727	2.0530	0.7587	1.0227	0.3620	0.9964	0.0323	0.0199

the interval τ, we require a higher regularization parameter λ_{reg} to obtain the best results. Additionally, we also observe that the best parameter fitting gets worse (in terms of MAD) the higher the τ, this is most likely due to information loss as we aggregate over counts over a coarser granularity. Similarly, note that the standard errors increase as well.

We conclude that our method produces estimations that are very close to the true values as compared to the MLE method, as illustrated in Table 3. Further, in contrast to the MLE method, our regularization technique only took 1.2 s (175 times faster than MLE) for parameter estimations, on average (Table 4).

Table 4 The learned parameters $\{a, \delta, \psi, \rho\}$ and their standard errors (enclosed in brackets) associated with the MLE method and our inference technique. For the proposed optimization method, the results with the best MAD in prediction correspond to each τ are shown. Further, the Hawkes–Cox process is shown to perform much better than the Homogeneous Poisson Process

Method	τ	λ_{reg}	a		δ		μ_{1G}		ρ		MAD
HPP	–	–	0.4334	0.0000	–	–	–	–	–	–	274
MLE	–	–	0.1876	0.0000	3.3782	0.0002	1.9156	0.0001	0.0001	0.0000	274
Functional Moment Matching Method	10	0.1	0.1003	0.0593	1.0208	1.5189	0.5721	0.7962	0.1547	0.1191	260
	30	0.1	0.0815	0.0820	1.6336	2.8913	1.0572	1.7038	0.0898	0.1083	207
	60	0.5	0.0654	0.0549	3.1429	3.8649	2.1338	2.4790	0.0594	0.0709	153
	120	0.3	0.0698	0.0621	2.1910	3.6380	1.5687	2.5068	0.0526	0.0803	183
	180	1.0	0.0690	0.0441	4.9340	3.7163	3.6692	2.6728	0.0117	0.0238	138
	360	**0.9**	**0.0428**	**0.0376**	**5.5690**	**4.7926**	**4.3617**	**3.6969**	**0.0084**	**0.0073**	**118**
	600	0.2	0.0420	0.0366	5.3028	4.8700	4.2769	3.9015	0.0161	0.0335	126

7 Disability Adjusted Life Years

DALY. The notion of the disability-adjusted life year (DALY) has been introduced in the 1990s. DALYs are important measurements that link the burden of disease in populations with the degree of morbidity, disability, and long-term survival. They were initially developed to gauge the global burden of disease and to determine a strategy for the evaluation of health benefits and its associated cost-effectiveness. One DALY is the equivalent of a year of healthy life lost if a person had not experienced a particular disease. Different theoretical and methodological challenges persist, which are likely to affect both the computation and interpretation of DALY estimates (Mathers et al. 2008).

The computation of DALY and its challenges. DALY calculations consist of two smaller estimates: (i) the number of years lived with disability (YLD); and (ii) the number of years of life lost (YLL) associated with the condition of interest. DALYs are estimated as the sum of the YLD and YLL pillars. YLD for different diseases is calculated using disease-specific disability weights that range between 0 (perfect health) and 1 (death) and the duration of disability. YLL is calculated using estimates of mortality associated with the condition of interest when untreated, life expectancy, and age at death.

At times, scarce, reliable population-based data on disease parameters, specifically incidence of deaths and diseases, make estimating DALY of a specific disease difficult, especially in lower middle-income economies (Kularatna et al. 2013; Wyber et al. 2015). Hence, standard techniques to deal with this uncertainty is to employ a sampling-based estimation procedure. In this approach, each parameter is modeled by a probability distribution centered upon the optimal estimate of the parameter with a range reflecting uncertainty (Puett et al. 2019; Noguera Zayas et al. 2021). Typically, disability weights were modeled using the continuous uniform distribution and the expected number of deaths were modeled using the Poisson distribution.

Proposed methodologies. Standard methods dictate the use of Poisson distribution to obtain estimates for the expected number of deaths and specific diseases. Recall that in a Poissonian framework, the probability of the number of deaths occurring in an age group happens with a fixed average rate, and is independent of the time since the last event. We propose the use of the intensity function in Eq. (2), for modeling purposes, which relaxes the Poisson assumption of stationary increments. It allows for the possibility that the rates need not be constant but can vary with time. This choice of intensity is general enough to encapsulate certain stylized facts of rates of deaths and diseases, such as nonlinear trend characteristics. Here, we illustrate the estimation procedure by applying the techniques presented earlier to obtain an estimate for the parameters for Hawkes–Cox process model, generalizing the Poissonian feature that is widely used in the literature. We first suppose that we observe the number of deaths (or number of diseases) across age groups, as tabulated in Table 5. For simplicity, we further suppose that there are three parameters to estimate, absorbed in $\theta \in \mathbb{R}^3$. Define the function $\mathcal{A}_\theta$ which takes the form

Table 5 Number of deaths or diseases, stratified by age

Age Group (↓), Year (→)	20xx	20xx+τ	20xx+2τ	...	20xx+2τ
0–4	R_1	R_2	R_3	...	R_N
5–14	...	...	...	...	...
15–44	...	...	...	...	...
45–59	...	...	...	...	...
60+	...	...	...	...	...

$$\mathcal{A}_\theta = \begin{pmatrix} m_1 - h_1^\tau \\ m_2 - h_2^\tau \\ m_3 - h_3^\tau \end{pmatrix}$$

where m_1, m_2, and m_3 denote the observed mean, variance, and covariance of the number of events within an interval with a predetermined specification of τ and δ given in (10), (11), and (12), respectively. The quantities h_1^τ, h_2^τ, and h_3^τ given in (6), (8), and (9), respectively, are functions of the parameters. Let R_i denote the number of deaths (or number of diseases) falling in age group i. Furthermore, let N be the number of age groups remaining. In this case, we have

$$m_1 = \frac{1}{N} \sum_{i=1}^{N} R_i \tag{10}$$

$$m_2 = \frac{1}{N} \sum_{i=1}^{N} R_i^2 - m_1^2 \tag{11}$$

$$m_3 = \frac{1}{N} \sum_{i=1}^{N} \left(R_i \times R_{i+\Delta} \right) - \left(\frac{1}{N} \sum_{i=1}^{N} R_i \right) \times \left(\frac{1}{N} \sum_{i=1}^{N} R_{i+\Delta} \right) \tag{12}$$

where $\Delta = \lceil \delta/\tau \rceil$. One can then evaluate $\hat{\theta}$ such that $\mathcal{A}_{\hat{\theta}} = 0$ using standard methods of moment matching (Hall 2004). With the estimated parameters, one can then obtain distributions related to the number of deaths and specific diseases. By a similar token, the same procedures may be applied to obtain suitable distributional properties for different age groups. This framework gives a natural way to deal with uncertainties in DALY computations, which subsumes the Poissonian assumptions.

8 Concluding Remarks

In this paper, we derive explicit moments and the covariance function of the number of events over an interval for the Hawkes–Cox process. The future evolution of this point process is influenced by the timing of past events whose intensities are affected by a self-exciting mechanism and an exogenous component. We develop an inference procedure through regularizations taking into account the theoretical functionals and showing that estimation is almost immediate. Empirical experiments on real data sets demonstrate its superior scalability and predictive performance.

References

Blundell, C., Beck, J., & Heller, K. A. (2012). Modelling reciprocating relationships with Hawkes processes. In *Advances in Neural Information Processing Systems*, (Vol. 25, pp. 2600–2608). Curran Associates.
Brémaud, P., & Massoulié, L. (1996). Stability of nonlinear Hawkes processes. *The Annals of Probability, 24*(3), 1563–1588.
Brémaud, P., & Massoulié, L. (2002). Power spectra of general shot noises and Hawkes point processes with a random excitation. *Advances in Applied Probability, 34*(1), 205–222.
Brémaud, P., Nappo, G., & Torrisi, G. L. (2002). Rate of convergence to equilibrium of marked Hawkes processes. *Journal of Applied Probability, 39*(1), 123–136.
Cox, D. R., & Isham, V. (1980). *Point Processes*. Chapman & Hall.
Cseke, B., & Heskes, T. (2011). Approximate marginals in latent Gaussian models. *Journal of Machine Learning Resesarch, 12*, 417–454.
Da Fonseca, J., & Zaatour, R. (2014). Hawkes process: fast calibration, application to trade clustering, and diffusive limit. *Journal of Futures Markets, 34*(6), 548–579.
Daley, D. J., & Vere-Jones, D. (2003). *An Introduction to the Theory of Point Processes, Elementary Theory and Methods*, (2nd ed., Vol. I). Springer.
Dassios, A., & Zhao, H. (2011). A dynamic contagion process. *Advances in Applied Probability, 43*(3), 814–846.
Davis, M. H. A. (1984). Piecewise-deterministic Markov processes: a general class of nondiffusion stochastic models. *Journal of the Royal Statistical Society. Series B. Methodological, 46*(3), 353–388.
Du, N., Wang, Y., He, N., and Song, L. (2015). Time-sensitive recommendation from recurrent user activities. In *Advances in Neural Information Processing Systems 28*, 3474–3482. Curran Associates.
Filimonov, V., & Sornette, D. (2012). Quantifying reflexivity in financial markets: Toward a prediction of flash crashes. *Physical Review E, 85*(5), 056108(1–9).
Fisher, N., & Banerjee, A. (2010). A novel kernel for learning a neuron model from spike train data. In *Advances in Neural Information Processing Systems 23*, 595–603. Curran Associates.
Gunter, T., Lloyd, C., Osborne, M. A., & Roberts, S. J. (2014). Efficient Bayesian nonparametric modelling of structured point processes. In *Uncertainty in Artificial Intelligence (UAI)*.
Hall, A. (2004). *Generalized Method of Moments*. Oxford University Press.
Hawkes, A. G. (1971). Spectra of some self-exciting and mutually exciting point processes. *Biometrika, 58*(1), 83–90.
Kularatna, S., Whitty, J. A., Johnson, N. W., & Scuffham, P. A. (2013). Health state valuation in low- and middle-income countries: A systematic review of the literature. *Value in Health, 16*(6), 1091–1099.

Kuss, M., & Rasmussen, C. E. (2005). Assessing approximate inference for binary Gaussian process classification. *Journal of Machine Learning Research, 6*, 1679–1704.

Linderman, S. W., & Adams, R. P. (2014). Discovering latent network structure in point process data. In *Thirty-First International Conference on Machine Learning (ICML)*.

Mangion, A. Z., Yuan, K., Kadirkamanathan, V., Niranjan, M., & Sanguinetti, G. (2011). Online variational inference for state-space models with point-process observations. *Neural Computation, 23*, 1967–1999.

Mathers, C., Fat, D. M., & Boerma, J. T. (2008). The global burden of disease: 2004 update.

Mohler, G. (2013). Modeling and estimation of multi-source clustering in crime and security data. *The Annals of Applied Statistics, 7*(3), 1525–1539.

Møller, J. (2003). Shot noise Cox processes. *Advances in Applied Probability, 35*(3), 614–640.

Møller, J., Syversveen, A. R., & Waagepetersen, R. P. (1998). Log Gaussian Cox processes. *Scandinavian Journal of Statistics, 25*(3), 451–482.

Nelder, J. A., & Mead, R. (1965). A simplex method for function minimization. *The Computer Journal, 7*(4), 308–313.

Noguera Zayas, L. P., Rüegg, S., & Torgerson, P. (2021). The burden of zoonoses in paraguay: A systematic review. *PLOS Neglected Tropical Diseases, 15*(11), 1–27.

Øksendal, B., & Sulem, A. (2007). *Applied Stochastic Control of Jump Diffusions* (2nd edn.). Springer.

Ozaki, T. (1979). Maximum likelihood estimation of Hawkes' self-exciting point processes. *Annals of the Institute of Statistical Mathematics, 31*(1), 145–155.

Puett, C., Bulti, A., & Myatt, M. (2019). Disability-adjusted life years for severe acute malnutrition: Implications of alternative model specifications. *Public Health Nutrition, 22*, 2729–2737.

Simma, A., & Jordan, M. I. (2010). Modeling events with cascades of poisson processes. In *UAI* (pp. 546–555). AUAI Press.

Teh, Y. W., & Rao, V. (2011). Gaussian process modulated renewal processes. In *Advances in Neural Information Processing Systems*, (Vol. 24, pp. 2474–2482). Curran Associates.

Wyber, R., Vaillancourt, S., Perry, W., Mannava, P., Folaranmi, T., & Celi, L. A. (2015). Big data in global health: Improving health in low- and middle-income countries. *Bulletin of the World Health Organization, 93*(3).

Young Lee AI research scientist at Roche Singapore, working in the fields of statistics and machine learning. Prior to this, he was a postdoctoral fellow in the Faculty of Arts and Sciences at Harvard University. Young holds a doctorate in Statistics from the London School of Economics.

Thanh Vinh Vo Research Fellow at the School of Computing, National University of Singapore (NUS). He completed his Ph.D. in computer science at the NUS on problems related to causal inference from observational data.

Derek Ni Data analyst lead in Roche Pharma International Informatics Data and Analytics Chapter, where he leads advanced analytics initiatives utilizing machine learning (ML), natural language processing (NLP) to generate key insights for go-to-market medical and commercial, leading to better patient care.

Gang Mu Holds a Ph.D. degree in mathematics. Comprehensive experiences to connect Mathematics, Healthcare and Technology together driving impacts and outcomes for patients and healthcare systems. Founded Swiss Network for Mathematics in Industry. Head of AI for Partnerships at Roche and Visiting Research Scholar at the University of Zurich.

Open Access This chapter is licensed under the terms of the Creative Commons Attribution 4.0 International License (http://creativecommons.org/licenses/by/4.0/), which permits use, sharing, adaptation, distribution and reproduction in any medium or format, as long as you give appropriate credit to the original author(s) and the source, provide a link to the Creative Commons license and indicate if changes were made.

The images or other third party material in this chapter are included in the chapter's Creative Commons license, unless indicated otherwise in a credit line to the material. If material is not included in the chapter's Creative Commons license and your intended use is not permitted by statutory regulation or exceeds the permitted use, you will need to obtain permission directly from the copyright holder.

The manufacturer's authorised representative in the EU is Springer Nature Customer Service Centre GmbH, Europaplatz 3, 69115 Heidelberg, Germany. If you have any concerns regarding our products, please contact ProductSafety@springernature.com

Printed and bound by CPI Group (UK) Ltd, Croydon, CR0 4YY

25/03/2026

02078170-0012